Indexed Dermal Bibliography
(1995–2007)

DEPARTMENT OF HEALTH AND HUMAN SERVICES
Centers for Disease Control and Prevention
National Institute for Occupational Safety and Health

> This document is in the public domain and may be freely copied or reprinted.

DISCLAIMER

Mention of any company or product does not constitute endorsement by the National Institute for Occupational Safety and Health (NIOSH). In addition, citations to Web sites external to NIOSH do not constitute NIOSH endorsement of the sponsoring organizations or their programs or products. Furthermore, NIOSH is not responsible for the content of these Web sites. All web addresses in this document were accessible as of the publication date.

ORDERING INFORMATION

To receive NIOSH documents or other information about occupational safety and health topics, contact NIOSH at

Telephone: **1–800–CDC–INFO** (1–800–232–4636)
TTY: 1–888–232–6348
E-mail: cdcinfo@cdc.gov

or visit the NIOSH Web site at **www.cdc.gov/niosh**

For a monthly update on news at NIOSH, subscribe to *NIOSH eNews* by visiting **www.cdc.gov/niosh/eNews**

DHHS (NIOSH) Publication Number 2009–153

September 2009

SAFER • HEALTHIER • PEOPLE™

FOREWORD

Workers from almost every industrial sector and trade routinely experience dermal exposures to chemicals via contact with contaminated surfaces, deposition of aerosols and vapors, and immersion in or splashes from liquids. Such exposures may result in adverse health consequences ranging from direct effects to the skin (e.g., irritant contact dermatitis and corrosion) to systemic effects (e.g., cancers and neurological effects) and to sensitization (e.g., allergic contact dermatitis). Occupational skin diseases have previously been identified as one of the leading causes of occupational illness within the United States workforce with many of the reported skin disorders being associated with chemical exposures.

The National Institute for Occupational Safety and Health (NIOSH) is dedicated to controlling and preventing workplace hazards including dermal exposures to chemicals. This document, *Indexed Dermal Bibliography (1995–2007)*, is intended to serve as a resource guide for information on dermal issues within the workplace. The *Indexed Dermal Bibliography* has been structured to accommodate varying levels of technical background or formal training in identifying and controlling harmful skin exposures. The primary topics covered within the *Indexed Dermal Bibliography* include: (1) an overview of dermal exposures, (2) hazard identification, (3) exposure characterization, (4) health effects surveillance, (5) risk assessment, and (6) risk control management. This resource guide is not designed to be an exhaustive compilation of materials from the dermal exposure literature, but rather a representative list of available dermal exposure resources. The *Indexed Dermal Bibliography* contains, for the most part, review articles and an overview of educational information.

Christine M. Branche, Ph.D. /s
Acting Director, National Institute
 for Occupational Safety and Health
Centers for Disease Control and Prevention

CONTENTS

Foreword v
Acknowledgments ix
Abbreviations and Acronyms xi
Definitions xix
Quick Guide to Using the Indexed Dermal Bibliography xxi

CHAPTER 1: Introduction 1

    1.1 Background 1
    1.2 Purpose of the Indexed Dermal Bibliography 1
    1.3 Intended Uses and Audiences 2
    1.4 Topics 2

CHAPTER 2: Indexed Dermal Bibliography Development and Format   3

    2.1 Criteria for Selection of Resources 3
    2.2 Resource Review Process 4
    2.3 Indexed Dermal Bibliography Contents 4
    2.4 Indexed Dermal Bibliography Format 5
    2.5 Obtaining Resources 6

CHAPTER 3: Resources for the General Audience 7

    3.1 Introduction 7
    3.2 Resources for the General Audience by Topic 7
            Topic 3A. Overview of Skin Exposures to Chemicals 9
            Topic 3B. Characterization of Exposure Condition
                    (Exposure Characterization) 13
            Topic 3C. Hazard Identification 15
            Topic 3D. Risk Assessment—Evaluating the Presence
                    of Harmful Chemicals 18
            Topic 3E. Risk Management—Skin Exposure
                    Risk Reductions 19

CHAPTER 4: Resources for the Professional Audience 25

    4.1 Introduction 25
    4.2 Resources for the Professional Audience by Topic 25

Contents

       Topic 4A. Overview of the Investigation and Control of Occupational Skin Exposures 27

       Topic 4B. Surveillance and Clinical Aspects 36

       Topic 4C. Exposure Characterization 41

       Topic 4D. Hazard Identification from Toxicological Studies or Modeling 53

       Topic 4E. Risk Assessment 66

       Topic 4F. Risk Management 69

CHAPTER 5: Overall Information Availability 75

Appendix A: Full Resource Citations and Summaries 79

ACKNOWLEDGMENTS

The *Indexed Dermal Bibliography* was developed in the Education and Information Division (EID), by Paul Schulte, Ph.D., Director; Chen-Peng Chen, Ph.D.;Thomas J. Lentz, Ph.D.; Dawn Tharr; and G. Scott Dotson, Ph.D. from material produced under contract # 200–2000–08017 with Westat. For contributions to the technical content and review of this document, the authors gratefully acknowledge the following NIOSH personnel:

    Heinz W. Ahlers, J.D., National Personal Protective Technology Laboratory (NPPTL), NIOSH

    Gregory A. Day, Ph.D., Division of Respiratory Disease Studies (DRDS), NIOSH

    Chad Dowell, Division of Surveillance, Hazard Evaluations and Field Studies (DSHEFS), NIOSH

    Charles L. Geraci, Ph.D., EID, NIOSH

    Cynthia Hines, DSHEFS, NIOSH

    Larry Reed, DSHEFS, NIOSH

    John P. Sestito, J.D., DSHEFS, NIOSH

    Sid Soderholm, Ph.D., Office of the Director (OD), NIOSH

    Douglas Trout, M.D., DSHEFS, NIOSH

The authors wish to thank Vanessa Becks, Gino Fazio, and Cathy Rotunda for their editorial support and contributions to the design and layout of this document.

Special appreciation is expressed to the following individuals and organizations for serving as independent external reviewers and providing comments that contributed to the development of this document:

    Ben Hayes, M.D., Ph.D., Adjunct Clinical Instructor, Division of Dermatology, Vanderbilt School of Medicine

    Sanja Kezic, Ph.D., Professor, Coronel Institute of Occupational Health, Academic Medical Center, University of Amsterdam

Acknowledgments

Youcheng Liu, M.D., Sc.D., Assistant Professor, Department of Preventive Medicine and Environmental Health, College of Public Health, University of Kentucky

Scott P. Schneider, Division Director, Occupational Safety and Health Division, Laborers' Health & Safety Fund of North America

James Taylor, M.D., Section Head, Industrial Dermatology, The Cleveland Clinic

ABBREVIATIONS AND ACRONYMS

AAFP	American Academy of Family Physicians
ACC	American Chemistry Council
ACGIH	American Conference of Governmental Industrial Hygienists
AFL-CIO	American Federation of Labor and Congress of Industrial Organizations
AID	allergic and irritant dermatitis
AIHA	American Industrial Hygiene Association
ANOM	analysis of means
ANSI	American National Standards Institute
API	Alliance for the Polyurethanes Industry
ASA	American Skin Association
ASTM	American Society for Testing and Materials
ATSDR	Agency for Toxic Substances and Disease Registry
BAT values	biological tolerance values
BEELs	biological environmental exposure limits
BEIs	biological exposure indices
BLS	Bureau of Labor Statistics
CA DIR	California Department of Industrial Relations
CAS	chemical abstract service
CBD	chronic beryllium disease
CCOHS	Canadian Centre for Occupational Health and Safety
CCP	carbonless copy paper
CDC	Centers for Disease Control and Prevention
CEB	Chemical Engineering Branch of the USEPA's OPPTS

Indexed Dermal Bibliography

Abbreviations and Acronyms

CEFIC	European Chemical Industry Council
CFD	computational fluid dynamics
CFR	Code of Federal Regulations
ChemIDplus	Free, web-based search system to access chemical substances cited in the National Library of Medicine databases
CICADs	Concise International Chemical Assessment Documents
CIS	International Occupational Safety and Health Information Centre (Centre international d'informations de sécurité et santé au travail)
CMAQ	Community Multiscale Air Quality Model
COSHH	Control of Substances Hazardous to Health
CPC	chemical protective clothing
CPFB	chloropentafluorobenzene
CPL	compliance directives
CPI	Center for the Polyurethane Industry
CPWR	Center for Construction Research and Training (Formerly known as the Center to Protect Workers' Rights)
CrVI	hexavalent chromium
DCB	dichlorobenzene
DEET	N,N-diethyl-m-toluamide
DERMDAT	dermal exposure measurements
DERP	Dermal Exposure Research Program
DFP	diisopropylfluorphosphate
DHHS	Department of Health and Human Services
DLI	Department of Labor and Industries

Abbreviations and Acronyms

DMF	N,N-dimethylformamide
DMSO	dimethylsulfoxide
DOEL	dermal occupational exposure limit
DP-PBPK	Distributed parameter-physiologically based pharmacokinetic models
DREAM	DeRmal Exposure AssessMent
EASE	Estimation Assessment of Substance Exposure mode
ECETOC	European Centre for Ecotoxicology and Toxicology of Chemicals
eLCOSH	Electronic Library of Construction Occupational Safety and Health
EPIDERM	Experience of the British dermatologists
ERDEM	Exposure-Related Dose-Estimating Model
ERG	Emergency Response Guidebook
ERPG	Emergency Response Planning Guidelines
EVOH	ethylene vinyl alcohol
EXTOXNET	EXTension TOXicology NETwork
FAQs	frequently asked questions
FDA	Food and Drug Administration
FIFRA	Federal Insecticide, Fungicide, and Rodenticide Act
FIOH	Finnish Institute of Occupational Health
FRAMES-3MRA	Framework for Risk Analysis in Multimedia Environmental Systems—Multimedia, Multipathway, Multireceptor Risk Assessment
FTIR	Fourier transform infrared spectroscopy
HSC	Health and Safety Commission
HSDB	Hazardous Substances Data Bank

HSE	Health and Safety Executive
IARC	International Agency for Research on Cancer
ICCVAM	Interagency Coordinating Committee on the Validation of Alternative Methods
ICSC	International Chemical Safety Card
ID	identifier
IDLHs	Immediately Dangerous to Life and Health values
ILO	International Labour Office
INRS	l'Institut National de Recherche et de Sécurité
IOM	Institute of Medicine
IPCS	International Programme on Chemical Safety
IPPSF	Isolated Perfused Porcine Skin Flap Model
IRIS	Integrated Risk Information System
ISEA	International Safety Equipment Association
IT	intrinsic toxicity
LAS	linear alkibenzene sulfonate
LLNA	local lymph node assay
MAK	maximum allowable concentration
MDA	4,4-methylene dianiline
MDI	Methylene diphenyl 4,4'-diisocyanate
MMA	methyl methacrylate
MMGs	medical management guidelines
NASD	National Ag Safety Database
NERL	The USEPA's National Exposure Research Laboratory
NIEHS	National Institute of Environmental Health Sciences
NIOSH	National Institute for Occupational Safety and Health

Abbreviations and Acronyms

NIOSHTIC	National Institute for Occupational Safety & Health Technical Information Center (database)
NLM	National Library of Medicine
NORA	National Occupational Research Agenda
NOSQ	Nordic Occupational Skin Questionnaire
NPL	National Priorities List
NRL	natural rubber latex
NRMCA	National Ready Mix Concrete Association
NTIS	National Technical Information Service
OCD	occupational contact dermatitis
OECD	Organisation for Economic Co-operation and Development
OELs	occupational exposure limits
OH	Ohio
OPPTS	USEPA Office of Prevention, Pesticides and Toxic Substances
OPRA	occupational physicians reporting activity
OR	Oregon
ORDHS	Oregon Department of Human Services
OR-OSHA	Oregon Occupational Safety and Health Division
OSH	occupational safety and health
OSHA	Occupational Safety and Health Administration
OWIIPP	Oregon Worker Illness and Injury Prevention Program
PAHs	polyaromatic hydrocarbons
PAR	provisional acceptable residues
PBPK	physiologically based pharmacokinetic

Abbreviations and Acronyms

PCBs	polychlorinated biphenyls
PDA	personal digital assistant
PDF	portable document format
PELs	permissible exposure limits
PHS	Public Health Service
PPE	personal protective equipment
ppm	part per million
PVA	polyvinylalcohol
PVC	polyvinylchloride
QSAR	quantitative structure-activity relationships
QSPRs	quantitative structure property relationships
RA	sulfate ricinolei acid
RAGS	Risk Assessment Guidance for Superfund
RCRA	Resource Conservation and Recovery Act
RD	low-level sulfur mustard
REACH	registration, evaluation, authorisation and restriction of chemical substances
RELs	recommended exposure limits
RISKofDERM	Risk assessment of occupational dermal exposure to chemicals
RTECS	Registry of Toxic Effects of Chemical Substances
SCT	Secretariat of Communications and Transportation of Mexico
SENSOR	Sentinel Event Notification System for Occupational Risks
SHARP	Safety and Health Assessment and Research for Prevention Program

SHEDS	Stochastic Human Exposure and Dose Simulation Model
SIS	Division of Specialized Information Services
SMSEs	small- and medium-sized enterprises
SRC	Syracuse Research Corporation
SRP	Scientific Review Panel
STEL	short-term exposure limit
TCDD	2,3,7,8-Tetrachlorodibenzo-p-Dioxin
TCE	trichloroethylene
TDI	Toluene diisocyanate
TEHIP	Toxicology and Environmental Health Information Program
TLVs	threshold limit values
TOXLINE	Toxicology Literature Online
TOXNET	Toxicology Data Network
TSCA	Toxic Substances Control Act
TSCATS	Toxic Substances Control Act Test Submission Database
TWA	time weighted average
USACHPPM	U.S. Army Center for Health Promotion and Preventive Medicine
UNEP	United Nations Environment Programme
USDOT	United Stated Department of Transportation
USEPA	United States Environmental Protection Agency
VOCs	volatile organic compounds
WADLI	Washington Department of Labor and Industry
WHO	World Health Organization

WISER	Wireless Information System for Emergency Responders
WEELs	Workplace Environmental Exposure Level handbook

DEFINITIONS

Article ID: The unique ID assigned to every resource found in the *Indexed Dermal Bibliography*.

Audiences: Whether the reference was written primarily for a general or professional audience. General audience is defined as those who have limited technical background or formal training in identifying and controlling hazardous skin exposures. Professional audiences typically utilize technical information for evaluating, recognizing, and controlling harmful skin exposures.

Chemical: The broad chemical classes—whether raw, intermediate, or final products—to which the resource pertains or specifically addresses (e.g., abrasives, pesticides, PCBs, etc.).

Citation: The information (e.g., author, title, journal, volume) needed to obtain the reference.

Educational material: Whether the material seem to have been developed with the primary focus of educating the workforce or general audiences.

Industries/Occupations: The broad categories of occupations and industries to which the resource states that it pertains or specifically addresses (e.g., agriculture, construction, mining).

Mixtures: Whether the references addresses the topic of chemical mixtures and dermal exposure.

Number of references: The number of references cited. NOTE: The total number for books was determined by summing the number of references in each chapter.

Resource type: Type of resource (e.g., journal, book, magazine, Web page). NOTE: Web sites may contain multiple types of resources not listed. These will be summarized in the summary text.

Definitions

Specific chemicals:	Specific chemicals to which the resource pertains or specifically addresses.
Specific process:	Specific occupations, jobs, or tasks, if any, addressed by the reference that are not listed or detailed in the industries/occupations drop-down list.
Summary:	A summary of the document written with a focus on occupational dermal exposures. The summary is not identical to the reference's abstract. Web site summaries may summarize links to resources within the Web site.
Topics addressed:	The list of broad topics addressed by the references (e.g., health effects, exposure characterization). Most resources contain information about multiple topics and subtopics. This list provides an overview of the kinds of information that can be found within the resource. Specific topics for general audiences are explained in full in Chapter 3 and for professional audiences in Chapter 4.

QUICK GUIDE TO USING THE *INDEXED DERMAL BIBLIOGRAPHY*

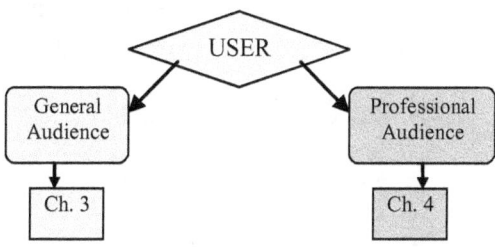

STEP 1. AUDIENCE

Select General or Professional Audience. (See Section 1.3 for definition of each audience.)

General Audience Tables	Professional Audience Tables
Table 3A. Overview of of skin exposures to chemicals	Table 4A. Overview of the investigation and control of occupational skin exposures
Table 3B. Characterization of exposure condition (exposure characterization)	Table 4B. Surveillance and clinical aspects
Table 3C. Hazard identification	Table 4C. Exposure characterization
Table 3D. Risk assessment—evaluating the presence of harmful Chemicals	Table 4D. Hazard identification from toxicological studies or modeling
Table 3E. Risk management—skin exposure risk Reduction	Table 4E. Risk assessment
	Table 4F. Risk management

STEP 2. TOPIC OF INTEREST

Select table with main topic of interest, as defined in Chapters 3 and 4.

Indexed Dermal Bibliography

Quick Guide to Using the Indexed Dermal Bibliography

STEP 3. RESOURCE TABLES

Go to Table and select resource(s) of interest based on:

1. Resource type
2. Title, Author
3. Year of publication
4. Sub-topic of interest, as defined in Chapters 3 and 4.
5. Article ID number for resource(s) of interest.

Proceed to Appendix A.

GENERAL AUDIENCE
Table 3A. Overview of skin exposures to chemicals

Resource type	ID	Title, author	Yr	Subtopics A.1	A.2	A.3
Book/monograph, whole	478	Essentials of Occupational Skin Management, PackhamCL	1999	✓	✓	✓
Brochure, pamphlet	107	Do you know about... the health hazards of benzene?,	2000	✓	✓	
	108	Do you know... the hazards of solvents?	2000	✓	✓	
Guideline	402	What You Need to Know About Occupational Exposure to Metal-working Fluids,	1998		✓	
Magazine article	2	Skin Care: Starting from Scratch, Nash,James L.	2000	✓		
	3	Dealing With Dermal Allergies and Skin Reactions, Groce,Don	2000	✓	✓	

STEP 4. APPENDIX A—FULL RESOURCE CITATION AND SUMMARY

In Appendix A, look up the ID numbers for resource(s) of interest. Resources are sorted numerically by ID number.

Evaluate the resource(s) based on the citation and summary information provided.

CHAPTER 1
Introduction

The *Indexed Dermal Bibliography (1995–2007)* is a tool that directs workers, employers, industrial hygienists, researchers, and policy makers to information resources on occupational skin exposures to chemicals, including health effects surveillance, exposure characterization, hazard identification, risk assessment, and risk control and management.

1.1 Background

More than 13 million workers in the United States are potentially exposed to chemicals at work via the skin. A worker's skin may be exposed to harmful chemicals through direct contact with contaminated surfaces, deposition of aerosols, and immersion in or splashes from liquids. Some chemicals cause contact dermatitis via direct skin contact. Contact dermatitis is one of the more frequently reported occupational illnesses, accounting for 10%–15% of all occupational diseases, at an estimated annual cost in the United States (U.S.) of at least $1 billion.

Many chemicals readily pass through the skin (called dermal penetration). Some of these chemicals are then taken up into the blood stream or by skin or immune cells (this is called dermal absorption). Dermal absorption can cause systemic health effects or can contribute to the effects of chemicals absorbed in the lungs by inhalation. Chemicals are often absorbed through the skin without being noticed by the worker. In some cases, the skin is a more significant route of exposure than the respiratory tract. This is particularly true for nonvolatile chemicals that are relatively toxic and that remain on work surfaces for long periods of time.

1.2 Purpose of the Dermal Resource Guide

The *Indexed Dermal Bibliography* is designed to serve as a resource for information on dermal exposure for those who work in (1) at-risk occupations, (2) positions to investigate or control worker skin exposure to harmful chemicals, and (3) research settings or positions to set policy on dermal exposures. The *Indexed Dermal Bibliography* provides lists and descriptions of resources by topic for people looking for specific information on dermal exposure anticipation, recognition, evaluation, and control.

The *Indexed Dermal Bibliography* is not designed to be an exhaustive listing of materials from the dermal exposure literature, but rather a representative list of available dermal exposure resources. The *Indexed Dermal Bibliography* contains review articles and summaries of educational information. Individual research studies are not included here. In addition, the accuracy of information presented in the references has not been evaluated.

1.3 Intended Uses and Audiences

The *Indexed Dermal Bibliography* is designed to provide descriptions of resources available for two different audiences: the general audience and the professional audience.

Resources for the general audience are for those who have limited technical background or formal training in identifying and controlling harmful skin exposures. The general audience may include workers, small business employers, supervisors, worksite owners, insurers, and manufacturers of industrial chemicals.

Resources for the professional audience are for those who typically use technical information for evaluating, recognizing, and controlling harmful skin exposures. The professional audience may include industrial hygienists, occupational epidemiologists, dermatologists, occupational physicians and nurses, academic researchers, toxicologists, and policy makers.

In some cases resources for these two audiences are not mutually exclusive. General audience members are encouraged to look at the professional resources when they are interested in more detailed or technical information. Conversely, professionals looking for background information for training, education, or communication purposes may find relevant information in the resources for the general audience.

1.4 Topics

The *Indexed Dermal Bibliography* lists resources that address a number of broad topics. The topics differ somewhat between general and professional audiences, but typically address the following:

- Overview of dermal exposure.
- Surveillance and clinical aspects of dermal exposures.
- Dermal exposure characterization.
- Dermal hazard identification.
- Dermal exposure risk assessment.
- Dermal exposure risk management.

These topics were specified during a workshop held at the International Conference on Occupational and Environmental Exposure of Skin to Chemicals: Science and Policy, held September 11, 2002 in Crystal City, VA. The topics are defined and discussed in more detail in Chapters 3 and 4.

CHAPTER 2
Indexed Dermal Bibliography Development and Format

2.1 Criteria for Selection of Resources

Resources listed in the *Indexed Dermal Bibliography* were identified by first conducting an electronic search of review articles on dermal exposure topics published in English between 1995 and 2007. Key words were identified, grouped, and used to search nine different bibliographic databases (Combined Health Information Database, Cumulative Index to Nursing and Allied Health Literature, Enviroline, Gale Group Health & Wellness Database^{SM}, Health Source—Consumer Edition, National Technical Information Service (NTIS), PubMed/Medline, Wilson Applied Science & Technology Abstracts, and WorldCat) and three search engines for government documents (FirstGov.gov, Google UncleSam.com, and SearchGov.com). Additional resources were identified from government guidelines, significant Web sites, and suggestions from dermal exposure experts. These efforts resulted in identification of over 600 resources potentially suitable for inclusion in this guide. Of these, 229 resources were retained in the final version of this guide.

In order to select the resources for inclusion in the *Indexed Dermal Bibliography*, reviewers first screened the abstracts and other available bibliographic information for each of the identified resources for relevance to the *Indexed Dermal Bibliography*. Based on the screening, each resource was either recommended, not recommended, or potentially recommended for inclusion in the guide. For a resource to be considered for inclusion, it needed to meet the following criteria:

- Be published in English between 1995 and 2007. (In rare instances, review articles or resources published before 1995 were selected for review when they were considered key resources.)
- Cover occupational exposures that are primarily via dermal pathways.
- Cover dermal exposures to chemical hazards, rather than physical or biological hazards.
- Deal primarily with human exposures or animal studies directly related to human exposures.
- Be nonclinical and designed for the nonphysician audience.
- Be a review article or a meta-analysis of primary research studies.

2.2 Resource Review Process

Following screening, copies of recommended resources were obtained and reviewed for inclusion in the *Indexed Dermal Bibliography*. All resources that were included in the Guide were reviewed more fully to summarize the content of the resource. For each of the references, the following information was gathered:

- Industry covered by the resources, including specific industries or occupations when they were discussed in-depth.
- Chemical classes covered in the resource, including any specific chemicals that were covered in-depth.
- Discussion of any issues surrounding dermal exposure to chemical mixtures.
- Audience—professional or general (see Section 1.3. above).
- Major topics associated with dermal exposure covered in the resources.

In addition, a brief summary of each resource was written, or, in the case of Web sites and Web pages, key information found within the resource was highlighted and as appropriate, summarized.

The draft *Indexed Dermal Bibliography* underwent an external review process which included review by the general public. Minor edits and additions were made in response to reviewer comments. Several newer articles were also added during the review process. Of the original 600+ resources, 229 resources met the previously described criteria and are included in the guide.

2.3 Indexed Dermal Bibliography Contents

The 229 resources in the *Indexed Dermal Bibliography* include review articles published in peer reviewed journals, occasional primary journal articles, books, book chapters, brochures/pamphlets, databases, government policies and regulations, guidelines, magazine articles, technical reports, and Web sites and Web pages. Individual research studies are not included here.

Web sites and Web pages were treated somewhat differently with respect to citations and summaries, given the unique nature of their content. First to clarify, Web sites contain a variety resources, including Web pages and downloadable documents, data files, and databases. Web sites are somewhat analogous to a book, with the Web site being the book and the Web pages being chapters within the book. However, Web sites are more complicated than books because Web pages can be much more varied in content and format than chapters. In addition, Web site and Web page content is not necessarily static content, as is the case with printed material. Information available on a Web site can change or be updated over time, can be removed, or can be moved to a different location within the Web site. It is not always clear on examination whether or not this has been done. Finally, the Web sites reviewed and included in the *Indexed Dermal Bibliography* contain a variety of independent information located throughout the Web sites as well as links to other Web sites. The advantage of resources available on Web sites is that they are instantaneously available and, for the most part, free, making Web sites an invaluable source of information for those addressing skin exposures in the field.

Web site summaries include general descriptions of Web pages, data files, and databases found within the Web site. The information cited in the *Indexed Dermal Bibliography* is based on the version of the Web site at the time it was reviewed.

2.4 Indexed Dermal Bibliography Format

The *Indexed Dermal Bibliography* is divided into six chapters and two appendices. The content of each is described below.

- Chapter 1: Background, purpose, target audiences, and topic areas.
- Chapter 2: *Indexed Dermal Bibliography* contents, format, and criteria for selection of resources.
- Chapter 3: General audience resources. The major topics covered are overview, exposure characterization, hazard identification, risk assessment, and risk management. Under each topic is a table listing those resources that include information on that topic, listed by resource type, ID number, author, date, and title.
- Chapter 4: Professional audience resources. The major topics covered are overview, surveillance and clinical aspects, exposure characterization, hazard identification, risk assessment, and risk management. Under each topic is a table listing those resources that include information on that topic, listed by resource type, ID number, author, date, and title.
- Chapter 5: Discussion of overall information availability.
- Appendix A: Full Resource Citations and Summaries. The appendix provides the citations and summaries of all resources in the *Indexed Dermal Bibliography*, sorted by resource

ID number. The Web site summaries list Web pages, databases, and data files found at the site which contain useful information on dermal exposures.

2.5 Obtaining Resources

Resources can be obtained through a variety of means. Books, journals, and magazine articles can be obtained through subscribing libraries or through interlibrary loan. As needed, books may also be purchased through a variety of suppliers.

More and more review articles are either available for purchase online or are free. One convenient source for finding articles available online is PubMed/Medline (http://www.ncbi.nlm.nih.gov/entrez/query.fcgi).

Some of the governmental and nongovernmental agency publications can also be found either for purchase or download online. Generally these can be located by typing in the report title using a search engine such as Google (www.google.com).

Web sites and Web pages can typically be accessed by the URL provided in Appendix A. If the URL link is no longer in use, Web pages or files may be searched by typing in key words on the Web site search engine. Alternatively, the table of contents links under the home page can be used. For Web resources described that do not have listed URLs, the resource might be found using the Web site search engine, typing in the name of the desired resource or keywords associated with the resource, and accessing the information. If the resource is still not found on the Web site or Web page, help can be solicited from the Web site masters, usually through a "help" or "contact us" Web page.

CHAPTER 3
Resources for the General Audience

3.1 Introduction

The resources identified below are appropriate for a general audience who wants background information on dermal exposures to chemicals. General audiences are those who desire to maintain a safe and healthful worksite, but have limited technical background or formal training in identifying and controlling hazardous skin exposures. The general audience may include workers, small business employers, supervisors, worksite owners, insurers, and manufacturers of industrial chemicals.

The resources presented in the tables are review articles published in peer-reviewed journals, as well as books, magazines, Web sites, regulatory guidelines, databases, brochures and other types of resources. These are not meant to be a comprehensive list of information available for the general audience, but rather a representative list of what is available. In addition, the accuracy of the information contained in any resource has not been evaluated.

General audience members who would like more detailed information on some of the topic areas are encouraged to also look at the resources identified for professional audiences in Chapter 4. For example, if a user would like more information on factors that influence exposure conditions (Table 3B, Subtopic B2), they could look in Chapter 4 at Table 4C under Subtopic C2, Description of Factors Influencing Exposure Conditions.

3.2 Resources for the General Audience by Topic

The following five tables list, for a general audience, resources covering each major topic related to occupational skin exposure to chemicals. The major topics are further divided into varying numbers of subtopics, each of which is represented in the columns on the right-hand side of the tables.

Descriptions of each topic and subtopics are provided before each table to assist users in deciding what kind of information they are interested in learning more about.

For the general audience, the five major topics and associated subtopics are:

Topic 3A. Overview of Skin Exposures to Chemicals
        A.1. Occurrence of Skin Exposures in the Workplace

Chapter 3: Resources for the General Audience

    A.2. Health Effects from Skin Exposures to Chemicals

    A.3. Dermal Regulations and Skin Notations

Topic 3B. Characterization of Exposure Condition (Exposure Characterization)

    B.1. Job/Tasks, Industries/Processes, or Chemicals with Skin Exposures

    B.2. Factors that Influence Exposure Conditions

    B.3. Protocols/Checklists to Characterize Exposure to Skin Hazards

Topic 3C. Hazard identification

    C.1. Risk Phrases, Hazard Symbols, Skin Designations

    C.2. Tables/Charts/Lists of Hazards for Specific Chemicals

    C.3. Protocols/Checklists to Identify Skin Hazards in the Workplace

Topic 3D. Risk Assessment—Evaluating the Presence of Harmful Chemicals

    D.1. Protocols/Checklists to Identify Exposure Risk

Topic 3E. Risk management—Skin Exposure Risk Reduction

    E.1. Overview of Skin Exposure Control Options

    E.2. Protocols/Checklists to Monitor Potential Exposures

    E.3. "Best practices"/Guidelines/Recommendations

    E.4. Guidelines/Recommendations for Post-exposure Skin Decontamination

Each of the five tables presented below include the following columns:

- Resource Type—book, brochure, journal article, Web site, etc.
- ID—unique number assigned to each resource and can be used to locate each resource in Appendix A. All resources in Appendix A are sorted alphabetically by resource type and then numerically by ID.
- Title, Author—title and author, if available, for each resource.
- Yr—the year of publication of the resource; for Web sites and Web pages, the year the Web site or Web page was reviewed for inclusion in the *Indexed Dermal Bibliography*.
- Subtopics—each subtopic addressed by a resource is checked in the appropriate column; subtopics are defined at the bottom of each table.

Chapter 3: Resources for the General Audience

A given resource may be repeated in multiple tables. This will happen when a resource provides information covering a variety of topics.

Topic 3A. Overview of Skin Exposures to Chemicals

Table 3A identifies resources that provide general background information on how skin exposure to chemicals in the workplace might cause health-related problems, as well as general information about health effects that can result from such exposures. Each checked box under subtopics indicates that that subject area is covered in the resource.

The following subtopics associated with exposure overview information are presented and defined below.

Subtopic A.1. Occurrence of Skin Exposures in the Workplace

These resources provide an overview of issues associated with the occurrence of skin exposures in the workplace.

Subtopic A.2. Health Effects from Skin Exposures to Chemicals

These resources contain general information on the health hazards associated with skin exposure to chemicals. Health effects from skin exposures to chemicals can vary ranging from local effects (e.g., irritation, burns, or skin breakdown) to allergic reactions, including both localized responses (e.g., hives) as well as more remote location responses (e.g., respiratory or lung effects).

Subtopic A.3. Dermal Regulations and Skin Notations

These resources contain information about regulations covering occupational skin exposure to chemicals. Unlike chemical inhalation hazards, there are currently no occupational exposure limits for skin exposures to chemicals. Instead, regulatory agencies assign a skin notation to a chemical to indicate that the chemical has the potential to contribute to the overall chemical exposure by absorption through the skin. Skin notations do not indicate whether a chemical can cause a localized response, only that skin exposure can contribute to overall exposure. The governmental Web sites listed contain information on applicable standards. In addition to regulatory information, some of these resources also contain lists of chemicals that have assigned skin notations.

Indexed Dermal Bibliography

Chapter 3: Resources for the General Audience

GENERAL AUDIENCE
Table 3A. Overview of skin exposures to chemicals

Resource type	ID	Title, author	Yr	Subtopics A.1*	A.2	A.3
Book/monograph, whole	169	Essentials of occupational skin management, Packham CL	1999	✓	✓	✓
Brochure, pamphlet	7	Do you know about... the health hazards of benzene? Occupational Health Department	2000	✓	✓	
	8	Did you know the hazards of solvents? Occupational Health Department	2000	✓	✓	
	9	Assessing and managing risks at work from skin exposure to chemical agents: Guidance for employers and health and safety specialists, Health and Safety Executive (HSE)	2001	✓	✓	✓
	12	Choice of skin care products for the workplace, HSE	2001		✓	
	24	2005 Emergency response planning guidelines (ERPG) and workplace environmental exposure level (WEEL) handbook, American Industrial Hygiene Association (AIHA)	2005			✓
	60	A safety and health practitioner's guide to skin protection, Center to Protect Workers' Rights (CPWR)	2000	✓	✓	✓
	62	An employer's guide to skin protection, CPWR	2005	✓	✓	
	95	Quick selection guide to chemical protective clothing (CPC), Forsberg K	2002			✓
Guideline	3	What you need to know about occupational exposure to metalworking fluids, National Institute for Occupational Safety and Health (NIOSH)	1998		✓	
Journal article—review, meta-analysis	43	Federal government regulation of occupational skin exposure in the USA, Boeniger MF	2003			✓
	147	The value and limitations of protective gloves in medical health service: Part III, Mellstrom GA	1996		✓	

*See footnotes at end of table.

(Continued)

GENERAL AUDIENCE
Table 3A (Continued). Overview of skin exposures to chemicals

Resource type	ID	Title, author	Yr	Subtopics A.1*	A.2	A.3
Magazine article	75	Protecting the hand-skin barrier in the workplace, Del Rosso J	2000		✓	
	149	Skin care: starting from scratch, Nash JL	2000	✓		
	181	Safe use of glutaraldehyde, Romano-Woodward D	2000		✓	
Web page	19	Toxicological profile information sheet, Agency for Toxic Substance and Disease Registry (ATSDR)	2005	✓	✓	✓
	21	Skin problems: How to protect yourself from job-related skin problems, American Academy of Family Physicians (AAFP)	2004	✓	✓	
	205	Health and safety zone, Unison	2005	✓	✓	
	220	Dermatitis: safety and health assessment and research for prevention (SHARP), Washington Department of Labour Industry (WADLI)	2005	✓	✓	
Web site	13	The American Skin Association, (ASA) [Home page], ASA	2005	✓	✓	
	14	Montana Department of Labor and Industries—Employment Relations, (MTDLI) [Home page], MTDLI	2005	✓	✓	
	15	Occupational health and safety, [Web site] 1105 Media, Inc.	2006	✓		✓
	16	Center for the Polyurethanes Industry (CPI) [Home page], American Chemistry Council (ACC)	2006	✓	✓	
	22	AAFP [Home page], AAFP	2005	✓	✓	
	29	Ansell Chemsafe [Home page], Ansell Chemsafe	2005	✓		

*See footnotes at end of table.

(Continued)

Chapter 3: Resources for the General Audience

GENERAL AUDIENCE
Table 3A (Continued). Overview of skin exposures to chemicals

Resource type	ID	Title, author	Yr	Subtopics A.1*	A.2	A.3
Web site (Continued)	58	Canadian Centre for Occupational Health and Safety (CCOHS) [Home page], CCOHS	2005	✓	✓	
	59	CPWR [Web page], CPWR	2006	✓		
	63	Electronic Library of Construction Occupational Safety and Health (eLCOSH) [Web site], CPWR	2005	✓	✓	
	88	European Agency for Safety and Health and Work [Home page], European Agency for Safety and Health and Work	2005	✓	✓	✓
	111	International Brotherhood of Teamsters [Home page], International Brotherhood of Teamsters	2006	✓	✓	
	112	International Labor Organization (ILO) [Home page], ILO	2005	✓	✓	✓
	156	National Ready Mixed Concrete Association (NRMCA) [Home page], NRMCA	2006	✓		
	162	Occupational Safety and Health Administration (OSHA) [Home page], OSHA	2005		✓	✓
	165	Oregon OSHA (OR-OSHA) [Home page], OR-OSHA	2006		✓	✓
	174	Portland Cement Association (PCA) [Home page], PCA	2006	✓	✓	
	207	United States Army Center for Health Promotion & Preventive Medicine (USACHPPM) [Home page], USACHPPM	2006	✓		

*A.1=Occurrence of Skin Exposures in the Workplace; A.2=Health Effects from Skin Exposures to Chemicals; A.3=Dermal Regulations and Skin Notations

Topic 3B. Characterization of Exposure Condition (Exposure Characterization)

Exposure characterization is the process of describing the qualities of a given environment that may influence exposure. These conditions may include the following:

- Source of the chemical.
- Amount of chemical a worker is exposed to, the amount of time a worker is exposed in a work day, and how often a worker is exposed in a given week.
- Routes of the exposure to the chemical, such as exposure through the skin, lungs, or through the mouth if food or drink is contaminated.
- Chemical and physical properties of the chemical.
- Work practices (i.e., or how work using the chemical is performed).

Subtopic B.1. Job/Tasks, Industries/Processes, or Chemicals with Skin Exposures

These resources may contain information on industries, processes, or jobs that are prone to expose workers to chemicals that are harmful to the skin. If available, the specific chemicals or classes of chemicals involved in the skin exposures are listed in the resource summary in Appendix A.

Subtopic B.2. Factors that Influence Exposure Conditions

These resources contain descriptions of factors that influence exposure conditions. Exposure conditions not only include the working conditions for a job being performed, but also the specific factors that influence exposure. Factors that can influence exposure conditions include the following: (1) intensity or amount of the exposure and (2) duration of exposure and frequency in a given day, week, month, or year. Other factors that influence exposure conditions include what control measures are in place to help reduce exposure, including engineering controls, work practices that either increase or decrease exposures, and the use of personal protective equipment (PPE) such as gloves. For example, two different workplaces with workers performing the same task can have different exposure conditions based on what kind of control measures are in use at each workplace.

Subtopic B.3. Protocols/Checklists to Characterize Exposure to Skin Hazards

The resources in this subtopic provide protocols and checklists that can be followed to characterize exposures to skin hazards. Only those resources with protocols or checklists specific to dermal exposure characterization are checked here.

GENERAL AUDIENCE
Table 3B. Characterization of exposure condition (exposure characterization)

Resource type	ID	Title, author	Yr	Subtopics B.1*	B.2	B.3
Book/monograph, whole	169	Essentials of occupational skin management, Packham CL	1999	✓	✓	
	185	Dermal exposure assessment, Sahmel J	2006	✓	✓	
Brochure, pamphlet	9	Assessing and managing risks at work from skin exposure to chemical agents: Guidance for employers and health and safety specialists, HSE	2001	✓		
	60	A safety and health practitioner's guide to skin protection, CPWR	2000	✓	✓	
	62	An employer's guide to skin protection, CPWR	2005	✓	✓	
Guideline	3	What you need to know about occupational exposure to metalworking fluids, NIOSH	1998		✓	
Web page	205	Health and safety zone [Home page], Unison	2005	✓		
	220	Dermatitis: Safety and health assessment and research for prevention (SHARP) [Home page], WADLI	2005	✓	✓	
Web site	14	MT DLI Employment Relations [Home page], MT DLI	2005	✓		
	15	Occupational health and safety [Web site], 1105 Media, Inc.	2006	✓		✓
	16	CPI [Web site], ACC	2006	✓	✓	
	22	AAFP [Home page], AAFP	2005	✓		
	58	CCOHS [Home page], CCOHS	2005	✓	✓	
	63	eLCOSH [Web site], CPWR	2005	✓	✓	✓

*See footnotes at end of table.

(Continued)

Chapter 3: Resources for the General Audience

GENERAL AUDIENCE
Table 3B (Continued). Characterization of exposure condition (exposure characterization)

				Subtopics		
Resource type	ID	Title, author	Yr	B.1*	B.2	B.3
Web site (Continued)	88	European Agency for Safety and Health and Work [Home page], European Agency for Safety and Health and Work	2005	✓		
	105	Skin at work, HSE	2005	✓	✓	✓
	111	International Brotherhood of Teamsters [Home page], International Brotherhood of Teamsters	2006	✓		
	112	ILO [Home page], ILO	2005	✓	✓	
	154	Toxicology Data Network(TOXNET)—Databases on toxicology, hazardous chemicals, environmental health, and toxic releases [Home page], National Library of Medicine (NLM)	2005	✓		
	162	OSHA [Home page], OSHA	2005	✓	✓	✓
	165	OR-OSHA [Home page], OR-OSHA	2006	✓	✓	
	174	PCA [Home page], PCA	2006		✓	
	207	USACHPPM [Home page], USACHPPM	2006	✓		✓

*B.1=Job/Tasks, Industries/Processes, or Chemicals with Skin Exposures; B.2=Factors that Influence Exposure Conditions; B.3=Protocols/Checklists to Characterize Exposure to Skin Hazards

Topic 3C. Hazard Identification

Hazard identification is the process of establishing the existence of a hazard through the existence of hazards through field observations and laboratory analysis of the exposures and adverse health effects. This includes the identification of chemical substances that are harmful to the skin or harmful to the body if absorbed through the skin. As part of this process, the nature of the hazard is determined, such as causes to skin irritation, skin corrosion, skin sensitization, or some type of effect elsewhere in the body from absorption through the skin. The resources listed in this table will help in determining the kind of skin hazards from different chemicals.

Subtopic C.1. Risk Phrases, Hazard Symbols, Skin Designations

These resources contain information on classifications of skin hazards associated with specific chemicals. Phrases, symbols, or other designations are used to describe the potential skin hazards. Skin hazards from chemicals can include (1) skin irritation and corrosion, (2) irritant contact dermatitis, (3) sensitization of skin and respiratory tract, (4) allergic contact dermatitis, (5) and contribution to overall body exposure.

Subtopic C.2. Tables/Charts/Lists of Hazards for Specific Chemicals

These resources include tables, charts, or lists of chemicals with the potential for significant skin exposures. These may include fact sheets that describe exposure conditions that may lead to harmful effects.

Subtopic C.3. Protocols/Checklists to Identify Skin Hazards in the Workplace

These resources provide protocols or checklists that can be used in the workplace to aid in the identification of skin hazards.

GENERAL AUDIENCE
Table 3C. Hazard identification

Resource type	ID	Title, author	Yr	C.1*	C.2	C.3
Book/monograph, whole	169	Essentials of occupational skin management, Packham CL	1999			✓
Brochure, pamphlet	7	Do you know about the health hazards of benzene? Occupational Health Department	2000	✓		
Brochure, pamphlet (Continued)	9	Assessing and managing risks at work from skin exposure to chemical agents: Guidance for employers and health and safety specialists, HSE	2001			✓
	95	Quick selection guide to CPC, Forsberg K	2002	✓	✓	
Technical publication/report	152	NIOSH Pocket Guide to Chemical Hazards, NIOSH	2004		✓	

*See footnotes at end of table.

(Continued)

Chapter 3: Resources for the General Audience

GENERAL AUDIENCE
Table 3C (Continued). Hazard identification

Resource Type	ID	Title, Author	Yr	Subtopics C.1*	C.2	C.3
Web page	151	International Chemical Safety Cards (ISCS): U.S. national version [Web site], NIOSH	2005	✓	✓	
	220	Dermatitis: Safety and health assessment and research for prevention (SHARP) [Home page], WADLI	2005		✓	
Web site	15	Occupational health and safety [Web site], 1105 Media, Inc.	2006	✓		✓
	22	AAFP [Home page], AAFP	2005		✓	
	58	CCOHS [Home page], CCOHS	2005			✓
	59	CPWR [Home page], CPWR	2006		✓	
	63	eLCOSH [Web site], CPWR	2005			✓
	88	European Agency for Safety and Health and Work [Home page], European Agency for Safety and Health and Work	2005		✓	
	105	Skin at work [Web site], HSE	2005			✓
	112	ILO [Home page], ILO	2005	✓	✓	✓
	154	TOXNET—Databases on toxicology, hazardous chemicals, environmental health, and toxic releases [Home page], NLM	2005	✓	✓	
	162	OSHA [Home page], OSHA	2005	✓	✓	✓

*C.1=Risk Phrases, Hazard Symbols, Skin Designations; C.2=Tables/Charts/Lists of Hazards for Specific Chemicals; C.3=Protocols/Checklists to Identify Skin Hazards in the Workplace

Indexed Dermal Bibliography

Chapter 3: Resources for the General Audience

Topic 3D. Risk Assessment—Evaluating the Presence of Harmful Chemicals

Risk Assessment is a measurement or an estimate of the chances of a given exposure to cause harm. With respect to skin exposures, risk assessments are performed by workplace health and safety representatives to provide the employer with some kind of estimate of the likelihood that an illness or injury will result from exposure of the skin to a particular chemical hazard. These resources provide guidance to identify whether a chemical skin exposure hazard is present in the workplace.

Subtopic D.1. Protocols/Checklists to Identify Exposure Risk

These resources provide protocols and checklists to be used to help individuals determine if an identified skin exposure hazard exists.

GENERAL AUDIENCE
Table 3D. Risk Assessment—Evaluating the presence of harmful chemicals

Resource type	ID	Title, author	Yr	Subtopics D.1*
Book/monograph, whole	169	Essentials of occupational skin management, Packham CL	1999	✓
	185	Dermal exposure assessments, Sahmel J	2006	✓
Brochure, pamphlet	60	A safety and health practitioner's guide to skin protection, CPWR	2000	✓
Web site	63	eLCOSH [Web site], CPWR	2005	✓
	105	Skin at work [Web site], HSE	2005	✓
	162	OSHA [Home page], OSHA	2005	✓

*D.1=Protocols/Checklists to Identify Exposure Risk

Topic 3E. Risk Management—Skin Exposure Risk Reduction

Risk management is the process of controlling risks to workplace hazards. These resources provide basic information on how to (1) monitor for potential exposures and (2) determine what control options are available, and (3) how to decontaminate skin once exposure has occurred.

Subtopic E.1. Overview of Skin Exposure Control Options

These resources provide an overview on how to control skin exposures to chemicals. There are a number of different kinds of controls available to minimize exposures of the skin to harmful chemicals.

- The most effective control approach is to eliminate the use of the harmful chemical or substitute a less harmful chemical in its place. An example would be replacing a solvent-based cleaner that causes skin drying and irritation with a water-based cleaner.
- If elimination or substitution is not possible, the next favored method of control is an engineering control. An example of an engineering control is enclosing a process that releases vapors or dusts that are irritating to the skin. The enclosure would remove the potential for contact with the skin. Another common engineering control is the use of ventilation, where an exhaust fan draws the chemical vapor or dusts from the work area.
- If engineering controls cannot be put in place, then work practices should be changed. This means changing the way a worker performs a job or task in order to lower exposures. For example, if skin contact is occurring through contaminated work surfaces, then work surfaces should be cleaned regularly or covered with a disposable material that can be replaced regularly.
- Administrative controls can also be used. These include training programs (e.g., programs that show workers how to avoid skin contact), hazard monitoring, and medical surveillance programs to determine if workers are being exposed to harmful chemicals.
- When all else fails, the use of personal protective equipment (PPE), such as gloves and coveralls, can be used to lower exposures to workers. It is important to remember that PPE only works if it is properly selected, properly put on and worn, and properly taken off. This means that workers must be trained regularly, have an adequate size selection and supply of PPE, and be monitored regularly to ensure proper fit and use.
- Finally, employers can also implement skin management programs that promote good skin care. This can include creams that provide a protective

barrier between the skin and the chemicals, as well as creams and lotions that remove chemicals from the skin and help maintain healthy skin.

Subtopic E.2. Protocols/Checklists to Monitor Potential Exposures

These resources contain protocols or checklists to use in routine qualitative monitoring of potential exposures to skin hazards. These will help identify whether chemical skin hazards are present in the workplace.

Subtopic E.3. "Best practices"/Guidelines/Recommendations

These resources include information on best practices, guidelines, or recommendations for chemical substitution, engineering controls, work practices, administrative controls, use of personal protective equipment, and implementation of a skin management program.

Subtopic E.4. Guidelines/Recommendations for Post-exposure Skin Decontamination

These resources contain information on how to decontaminate skin once skin exposures have occurred.

GENERAL AUDIENCE
Table 3E. Risk management—skin exposure risk reduction

Resource type	ID	Title, author	Yr	E.1*	E.2	E.3	E.4
Book/monograph, whole	169	Essentials of occupational skin management, Packham CL	1999	✓	✓	✓	✓
Brochure, pamphlet	6	Selecting protective gloves for work with chemicals, HSE	2000			✓	
	7	Do you know about the health hazards of benzene?, Occupational Health Department	2000			✓	
	8	Did you know the hazards of solvents?, Occupational Health Department	2000			✓	

*See footnotes at end of table.

(Continued)

Chapter 3: Resources for the General Audience

GENERAL AUDIENCE
Table 3E (Continued). Risk management—Skin Exposure Risk Reduction

Resource type	ID	Title, author	Yr	Subtopics			
				E.1*	E.2	E.3	E.4
Brochure, pamphlet (Continued)	9	Assessing and managing risks at work from skin exposure to chemical agents: Guidance for employers and health and safety specialists, HSE	2001			✓	
	11	Cost and effectiveness of chemical protective gloves for the workplace, HSE	2001			✓	
	12	Choice of skin care products for the workplace, HSE	2001			✓	
	60	A safety and health practitioner's guide to skin protection, CPWR	2000	✓	✓	✓	✓
	62	An employer's guide to skin protection, CPWR	2005	✓		✓	✓
	95	Quick selection guide to CPC, Forsberg K	2002			✓	
Guideline	3	What you need to know about occupational exposure to metalworking fluids, NIOSH	1998			✓	
Journal article—review, meta-analysis	145	Systemic toxicity from skin exposures (or what happens when you do not decontaminate), McDougal JN	2007				✓
	147	The value and limitations of protective gloves in medical health service: Part III, Mellstrom GA	1996			✓	
Magazine article	52	Chemical hand protection, Brown JW	2002			✓	
	72	Helping hands. Skin care for the hands, Crassweller I	1999			✓	✓
	75	Protecting the hand-skin barrier in the workplace, Del Rosso, J	2001			✓	

*See footnotes at end of table.

(Continued)

GENERAL AUDIENCE
Table 3E (Continued). Risk management—Skin Exposure Risk Reduction

Resource type	ID	Title, author	Yr	Subtopics			
				E.1*	E.2	E.3	E.4
Magazine article (Continued)	149	Skin care: Starting from scratch, Nash JL	2000	✓		✓	
	181	Safe use of glutaraldehyde, Romano-Woodward D	2000		✓	✓	✓
	186	Protecting hands against chemical exposures, Sarkis K	2000			✓	
Other—Guideline from private lab	70	A guide to dermal exposure reduction, Color-metric Laboratories Inc.	1999	✓			
Technical publication/report	152	NIOSH pocket guide to chemical hazards, NIOSH	2004			✓	✓
Web page	21	Skin problems: How to protect yourself from job-related skin problems [Web site], AAFP	2004	✓			
	150	Recommendations for CPC: A companion to the NIOSH pocket guide to chemical hazards [Web site], NIOSH	2005			✓	
	151	ISCS: U.S. national version [Web site], NIOSH	2005	✓		✓	✓
	205	Health and safety zone [Home page], Unison	2005	✓			
	220	Dermatitis: safety and health assessment and research for prevention (SHARP) [Home page] WADLI	2005	✓		✓	
Web site	14	MT DLI—Employment Relations [Home page], MT DLI	2005	✓		✓	✓
	15	Occupational safety and health [Web site], 1105 Media, Inc.	2006		✓	✓	
	16	CPI, [Web site], ACC	2006	✓		✓	✓
	22	AAFP [Home page], AAFP	2005	✓		✓	

*See footnotes at end of table.

(Continued)

GENERAL AUDIENCE

Table 3E (Continued). Risk management—Skin Exposure Risk Reduction

Resource type	ID	Title, author	Yr	Subtopics			
				E.1*	E.2	E.3	E.4
Web site (Continued)	29	Ansell Chemsafe [Home page], Ansell Chemsafe	2005	✓		✓	
	58	CCOHS [Home page], CCOHS	2005			✓	
	59	CPWR [Home page], CPWR	2006	✓		✓	✓
	63	eLCOSH [Web site], CPWR	2005			✓	✓
	64	National Ag Safety Database (NASD) [Web site], Centers for Disease Control and Prevention [CDC]	2006	✓		✓	✓
	88	European Agency for Safety and Health and Work [Home page], European Agency for Safety and Health and Work	2005			✓	
	105	Skin at work [Web site], HSE	2005	✓	✓	✓	
	111	International Brotherhood of Teamsters [Home page], International Brotherhood of Teamsters	2006	✓			
	112	ILO [Home page], ILO	2005	✓		✓	✓
	154	TOXNET—Databases on toxicology, hazardous chemicals, environmental health, and toxic releases [Home page], NLM	2005	✓			✓
	156	National Ready Mixed Concrete Association [Home page], NRMCA	2006	✓		✓	
	162	OSHA [Home page], OSHA	2005	✓	✓	✓	✓
	165	OR-OSHA [Home page], OR-OSHA	2006	✓		✓	
	174	PCA [Home page], PCA	2006	✓		✓	

*See footnotes at end of table.

(Continued)

GENERAL AUDIENCE
Table 3E (Continued). Risk management—Skin Exposure Risk Reduction

				Subtopics			
Resource type	ID	Title, author	Yr	E.1*	E.2	E.3	E.4
Web site (Continued)	207	USACHPPM [Home page], USACHPPM	2006			✓	✓
	225	International Programme on Chemical Safety (IPCS) [Home page], WHO	2005				✓

*E.1=Overview of Skin Exposure Control Options; E.2=Protocols/Checklists to Monitor Potential Exposures; E.3="Best Practices"/Guidelines/Recommendations; E.4=Guidelines/Recommendations for Postexposure Skin Decontamination

CHAPTER 4
Resources for the Professional Audience

4.1 Introduction

The resources identified below are appropriate for a professional audience to use in investigating and controlling harmful skin exposures in the workplace. Included in the professional audience are those who may be involved in conducting scientific risk assessments and preparing technical recommendations. In general, they have adequate knowledge to use technical information for evaluating, recognizing, and controlling harmful skin exposures. The professional audience may include industrial hygienists, occupational epidemiologists, dermatologists, occupational physicians and nurses, academic researchers and policy makers.

Professional audience members who are looking for more general treatments of some of the topics provided below, including material appropriate for worker training activities, are encouraged to also look at the resources identified for the general audience in Chapter 3.

As a reminder, the resources are review articles published in peer reviewed journals, as well as books, magazines, Web sites, regulatory guidelines, databases, brochures and a number of other types of resources. Also, they are not meant to be a comprehensive list of review information available for the professional audience, but rather a representative list of what is available. In addition, the accuracy of the information contained in any resource has not been evaluated.

4.2 Resources for the Professional Audience by Topic

The following tables provide a review of resources covering each of six major topics, as listed below, related to occupational skin exposure to chemicals for a professional audience. The major topics are further divided into subtopics, each of which is represented in the columns on the right-hand side of the tables. Descriptions of each topic and its related subtopics are given before each table to assist users in deciding what kind of information they are interested in obtaining.

For the professional audience, the six major topics and related subtopics are:

    Topic 4A. Overview of the Investigation and Control of Occupational Skin Exposures

Chapter 4: Resources for the Professional Audience

- A.1. Occurrence of Skin Exposures in the Workplace
- A.2. Health Hazards Resulting from Skin Exposure to Chemicals
- A.3. Investigation, Intervention, and Control of Occupational Skin Exposures
- A.4. Skin Physiology and Function as Barriers to Chemical Insults
- A.5. Dermal Regulations and Skin Notations

Topic 4B. Surveillance and Clinical Aspects
- B.1. Surveillance Studies Reporting Incidences of Occupational Skin Exposures
- B.2. Loss of Workdays and Impact on Productivity
- B.3. Surveillance Study Protocols/Procedures for Gathering Data
- B.4. Clinical Protocols for Recognition of Skin Exposure Health Effects

Topic 4C. Exposure Characterization
- C.1. Workplace Factors Associated with Harmful Skin Exposures
- C.2. Description of Factors Influencing Exposure Conditions
- C.3. Checklists/Questionnaires to Quantify Skin Exposure Incidents
- C4. Methods to Measure Exposures
- C.5. Exposure Modeling

Topic 4D. Hazard Identification from Toxicological Studies or Modeling
- D.1. Potential Health Effects Resulting from Specific Chemicals
- D.2. Summaries of Health Effects and Dose-Response Relationships
- D.3. Characterization Protocols

Topic 4E. Risk Assessment
- E.1. Guidelines for Risk Assessment or Analysis
- E.2. Example of Risk Assessment

Topic 4F. Risk Management
- F.1. Exposure Control Strategies
- F.2. Risk Assessment Protocols

Chapter 4: Resources for the Professional Audience

As in Chapter 3, each of the tables presented here include the following columns:

- Resource Type—whether the resource is a book, brochure, journal article, Web site, and so forth.
- ID—unique ID number assigned to each resource, can be used to locate each resource in Appendix A, organized in alphabetical/numerical order by resource type.
- Title, Author—the resource title and author, if listed, for each resource.
- Yr—the publication year of the resource and, in the case of Web sites and Web pages, the year the Web site or Web page was reviewed for inclusion in the *Indexed Dermal Bibliography*.
- Subtopics—Subtopics addressed under the given table topic. The definition of each subtopic is listed at the bottom of each table. Each subtopic addressed by a resource is checked.

A given resource may be repeated in multiple tables. This will happen when a resource provides information covering a variety topic areas.

Topic 4A. Overview of the Investigation and Control of Occupational Skin Exposures

These resources provide an overview of the occurrence of skin exposures to chemicals in the workplace; health hazards that can result from skin exposure to chemicals; the process of investigating, intervening and controlling occupational skin exposures; and background information on basic skin physiology and how skin functions as a barrier to chemical absorption into the body.

Subtopic A.1. Occurrence of Skin Exposures in the Workplace

These resources provide a general overview of the occurrence of skin exposures to harmful chemicals in the workplace. This overview may not be the primary focus of the resource, but rather introductory material that leads into the main focus of the resource.

Subtopic A.2. Health Hazards Resulting from Skin Exposure to Chemicals

These resources provide an overview description of the different kinds of adverse health effects that can result from skin exposure to chemicals.

Indexed Dermal Bibliography 27

Subtopic A.3. Investigation, Intervention, and Control of Occupational Skin Exposures

These resources provide an introduction to the recognition, evaluation and control of skin hazards in the workplace. Background information can be found on skin hazards, including an overview of occupational skin exposure investigations, intervention approaches that can be adopted and control strategies that can be put into place.

Subtopic A.4. Skin Physiology and Function as a Barrier to Chemical Insults

These resources provide an introduction to skin physiology, including the function of the different layers of skin, how they act as barries to chemical absorption, and how they can alter the skin's natural barrier properties when compromised or damaged.

Subtopic A.5. Dermal Regulations and Skin Notations

These resources either contain information about regulations covering occupational skin exposure to chemicals or information about the designation of chemicals based on their skin hazard or their potential to be absorbed by the skin. The governmental Web sites listed contain information on applicable standards. In addition to regulatory information, some of these resources also contain information on chemicals with skin notations.

PROFESSIONAL AUDIENCE
Table 4A. Overview of the investigation and control of occupational skin exposures

Resource type	ID	Title, author	Yr	A.1*	A.2	A.3	A.4	A.5
Book/monograph, chapter	4	Occupational skin exposure-absorption of chemical agents and assessment of exposures, Harris R	2000				✓	
	28	Systemic toxicity from percutaneous absorption, Andersen KE	1999				✓	

*See footnotes at end of table.

(Continued)

PROFESSIONAL AUDIENCE
Table 4A (Continued). Overview of the investigation and control of occupational skin exposures

Resource type	ID	Title, author	Yr	A.1*	A.2	A.3	A.4	A.5
Book/monograph, chapter	40	CPC and the skin: Practical considerations, Boeniger M	2002				✓	
	45	Protective gloves for occupational use, Boman A	2005					✓
	157	Surface and dermal monitoring for toxic exposures, Ness SA	1994		✓	✓		
	185	Dermal exposure assessments, Sahmel J	2006				✓	
	218	Health risk assessment: Dermal and inhalation exposure and absorption of toxicants (dermatology), Wang RGM	1993				✓	
	221	Development of occupational skin disease, Weber LW	2003				✓	
	229	Dermatotoxicology, Zhai H	2004				✓	
Brochure, pamphlet	60	A safety and health practitioner's guide to skin protection, CPWR	2000	✓	✓		✓	
	61	Physician's alert for occupational contact dermatitis among construction workers, CPWR	2001	✓	✓			
Guideline	47	Guideline for hand hygiene in healthcare settings, CDC	2002				✓	
	161	Sampling for surface contamination, OSHA	2005					✓
Journal article—primary	37	Slow curing of aliphatic polyisocyanate paints in automotive refinishing: A potential source for skin exposure, Bello D	2007	✓				

*See footnotes at end of table.

(Continued)

PROFESSIONAL AUDIENCE
Table 4A (Continued). Overview of the investigation and control of occupational skin exposures

Resource type	ID	Title, author	Yr	Subtopics				
				A.1*	A.2	A.3	A.4	A.5
Journal article—primary	68	Surveillance of occupational skin disease: EPIDERM and OPRA, Cherry N	2000	✓				
	123	Skin cleansers for occupational use: testing the skin compatibility of different formulations, Klotz A	2003				✓	
	131	Skin exposure to aliphatic polyisocyanates in the auto body repair and refinishing industry: A qualitative assessment, Liu Y	2007	✓				
	142	Dermal exposure and urinary 1-hydroxypyrene among asphalt roofing workers, McClean MD	2007	✓				
Journal article—review, meta-analysis	1	Skin lesions and environmental exposures. An overview for the occupational health nurse, ATSDR	1996		✓		✓	
	30	Occupational contact dermatitis, Antezana, M	2003		✓			
	34	Classification criteria for skin-sensitizing chemicals: A commentary, Basketter DA	1999			✓		✓
	35	Factors affecting thresholds in allergic contact dermatitis: Safety and regulatory considerations, Basketter DA	2002				✓	
	38	Skin exposure to isocyanates: Reasons for concern, Bello D	2007	✓				

*See footnotes at end of table.

(Continued)

PROFESSIONAL AUDIENCE
Table 4A (Continued). Overview of the investigation and control of occupational skin exposures

Resource type	ID	Title, author	Yr	Subtopics				
				A.1*	A.2	A.3	A.4	A.5
Journal article— review, meta- analysis (Continued)	46	Proposal for the assessment of quantitative dermal exposure limits in occupational environments: Part 1. Development of a concept to derive a quantitative dermal occupational exposure limit, Bos PM	1998					✓
	51	Concepts of skin protection: Considerations for the evaluation and terminology of the performance of skin protective equipment, Brouwer DH	2005			✓	✓	
	69	Occupational issues of irritant contact dermatitis, Chew AL	2003	✓				
	73	Pesticide-related illness among migrant farm workers in the United States, Das R	2001	✓				
	78	The epidemiology of occupational contact dermatitis, Diepgen TL	1999	✓				
	79	Skin-conditioning products in occupational dermatology, Elsner P	2003				✓	
	81	Occupational contact dermatitis II: Risk assessment and prognosis, Emmett EA	2003			✓	✓	
	83	Detailed review document on classification systems for skin irritation/corrosion in OECD member countries, OECD	1999					✓

*See footnotes at end of table.

(Continued)

PROFESSIONAL AUDIENCE
Table 4A (Continued). Overview of the investigation and control of occupational skin exposures

Resource type	ID	Title, author	Yr	A.1*	A.2	A.3	A.4	A.5
Journal article—review, meta-analysis (Continued)	97	Developing control of substances hazardous to health (COSHH) regulations essentials: Dermal exposure, personal protective equipment and first aid, Garrod AN	2003					✓
	101	CCP and workplace safety: A review, Graves CG	2000	✓	✓			
	103	Occupation-related allergies in dentistry, Hamann CP	2005	✓	✓			
	106	Misinterpretation and misuse of exposure limits, Hewett P	2001					✓
	115	Dermal absorption of benzene: Implications for work practices and regulations, Kalnas J	2000	✓				✓
	118	The role of the skin in the development of chemical respiratory hypersensitivity, Kimber I	1996		✓			
	122	A critique of assumptions about selecting chemical-resistant gloves: A case for workplace evaluation of glove efficacy, Klingner TD	2002					✓
	124	Occupational contact dermatitis. Recognition and management, Koch P	2001		✓			
	134	The importance of occupational skin diseases in the United States, Lushniak BD	2003	✓				
	135	Occupational contact dermatitis, Lushniak BD	2004	✓	✓	✓		

*See footnotes at end of table.

(Continued)

PROFESSIONAL AUDIENCE
Table 4A (Continued). Overview of the investigation and control of occupational skin exposures

Resource type	ID	Title, author	Yr	Subtopics				
				A.1*	A.2	A.3	A.4	A.5
Journal article—review, meta-analysis (Continued)	138	Harmonization of future needs for dermal exposure assessment and modeling: a workshop report, Marquart H	2001					✓
	140	Dermal toxicity due to industrial chemicals, Mathur AK	2002				✓	
	144	Methods for assessing risks of dermal exposures in the workplace, McDougal JN	2002			✓	✓	
	159	Criteria for skin notation in different countries, Nielsen JB	2004					✓
	183	Solvents and the skin, Rowse DH	2004		✓		✓	✓
	187	Percutaneous penetration studies for risk assessment, Sartorelli P	2000			✓		
	188	Dermal exposure assessment in occupational medicine, Sartorelli P	2002	✓				
	192	When should a substance be designated as sensitizing for the skin ('Sh') or for the airways ('Sa')? Schnuch A	2002					✓
	194	Dermal exposure to chemicals in the workplace: Just how important is skin absorption? Semple S	2004	✓	✓		✓	
	197	From xenobiotic chemistry and metabolism to better prediction and risk assessment of skin allergy, Smith Pease CK	2003				✓	
	203	Management of dermatitis in the rubber manufacturing industry, Toeppen-Sprigg B	1999	✓	✓	✓		

*See footnotes at end of table.

(Continued)

Chapter 4: Resources for the Professional Audience

PROFESSIONAL AUDIENCE
Table 4A (Continued). Overview of the investigation and control of occupational skin exposures

| Resource type | ID | Title, author | Yr | Subtopics |||||
				A.1*	A.2	A.3	A.4	A.5
Journal article—review, meta-analysis (Continued)	204	Prediction of irritancy in the human skin irritancy model and occupational setting, Tupker RA	2003				✓	
	210	From dermal exposure to internal dose, van de Sandt J	2008				✓	
	213	Review of skin permeation hazard of bitumen fumes, van Rooij JG	2008	✓	✓	✓	✓	
Other—commentary	41	Exposure and absorption of hazardous materials through the skin, Boeniger MF	2000					✓
Other—editorial	92	Dermal exposure: A decade of real progress, Fenske RA	2000			✓		
Technical publication/report	114	European Chemical Industry Council (CEFIC) Workshop on methods to determine dermal permeation for human risk assessment, Institute of Medicine (IOM)	2004				✓	
	129	Epidemiology of skin and respiratory diseases among hairdressers, FIOH	2001	✓	✓			
	152	NIOSH pocket guide to chemical hazards, NIOSH	2004					✓
	226	IPCS: Environmental health criteria document on dermal absorption, WHO	2005				✓	
Web page	19	Toxicological profile information sheet, ATSDR	2005		✓	✓		
	155	Hazardous Substances Data Bank (HSDB) [Web site], NLM	2005					✓

*See footnotes at end of table.

(Continued)

PROFESSIONAL AUDIENCE
Table 4A (Continued). Overview of the investigation and control of occupational skin exposures

Resource type	ID	Title, author	Yr	A.1*	A.2	A.3	A.4	A.5
Web page (Continued)	220	Dermatitis: Safety and health assessment and research for prevention (SHARP) [Home page], WADLI	2005	✓		✓	✓	
Web site	16	CPI [Web site], ACC	2006		✓			
	18	ATSDR [Home page], ATSDR	2005	✓				
	22	AAFP [Home page], AAFP	2005	✓	✓			
	25	AIHA [Home page], AIHA	2006					✓
	82	Dermatological engineering [Web site], Enviroderm Services	2005	✓	✓		✓	
	105	Skin at work [Web site], HSE	2005	✓	✓	✓	✓	
	112	ILO [Home page], ILO	2005	✓	✓	✓	✓	✓
	153	NIOSH [Home page], NIOSH	2005	✓	✓	✓		✓
	154	TOXNET—Databases on toxicology, hazardous chemicals, environmental health, and toxic releases [Home page], NLM	2005	✓	✓	✓		✓
	162	OSHA [Home page], OSHA	2005	✓	✓			✓
	206	The Extension Toxicology Network (EXTOXNET) [Web site], Ohio State University	2006		✓			✓

*A.1=Occurrence of Skin Exposures in Workplace; A.2=Health Hazards Resulting from Skin Exposure to Chemicals; A.3=Investigation, Intervention, and Control of Occupational Skin Exposures; A.4=Skin Physiology and Function as a Barrier to Chemical Insults; A.5=Dermal Regulations and Skin Notations

Topic 4B. Surveillance and Clinical Aspects

Public health surveillance is the ongoing systematic collection, analysis, and interpretation of health data for purposes of improving health and safety. *Occupational surveillance* studies involve the tracking of illnesses, injuries, exposures and hazards in the workplace. The resources provided in the following table review information on dermal exposure surveillance work as well as provide clinical guidance on collecting data for surveillance studies, including information on assessing health effects from skin exposure to chemicals in the workplace.

Subtopic B.1. Surveillance Studies Reporting Incidences of Occupational Skin Exposures

These resources summarize or refer to surveillance studies that report the incidence or prevalence of occupational skin exposures to chemicals. These may be studies where skin exposure to chemicals was a major focus of the study, or a minor focus of the surveillance study.

Subtopic B.2. Loss of Workdays and Impact on Productivity

These resources review information on lost workdays due to skin exposure health effects or review information on the impact of skin exposure health effects on worker productivity. This may include information on workers' compensation claims associated with skin exposure health effects.

Subtopic B.3. Surveillance Study Protocols/Procedures for Gathering Data

These resources contain guidance for collecting representative data for surveillance studies, including standard protocols or prevailing procedures used in surveillance studies.

Subtopic B.4. Clinical Protocols for Recognition of Skin Exposure Health Effects

These resources contain guidance for standard protocols or prevailing procedures used in clinical examinations that facilitate the recognition and identification of harmful health effects that result from skin exposures to chemicals.

Chapter 4: Resources for the Professional Audience

PROFESSIONAL AUDIENCE
Table 4B. Surveillance and clinical aspects

				Subtopics			
Resource type	ID	Title, author	Yr	B.1*	B.2	B.3	B.4
Book/monograph, chapter	28	Systemic toxicity from percutaneous absorption, Andersen KE	1999	✓			
	40	Chemical protective clothing and the skin: Practical considerations, Boeniger M	2002	✓			
	221	Development of occupational skin disease, Weber LW	2003				✓
Book/monograph, whole	137	Contact and occupational dermatology, Marks JG	2002				✓
Brochure, pamphlet	60	A safety and health practitioner's guide to skin protection, CPWR	2000	✓	✓	✓	✓
	61	Physician's alert for occupational contact dermatitis among construction workers, CPWR	2001				✓
Journal article—primary	55	Occupational dermatitis causing days away from work in U.S. private industry, 1993, Burnett CA	1998	✓	✓	✓	
	68	Surveillance of occupational skin disease: EPIDERM and OPRA, Cherry N	2000	✓			
	131	Skin exposure to aliphatic polyisocyanates in the auto body repair and refinishing industry: A qualitative assessment, Liu Y	2007	✓			
	196	Occupational contact dermatitis to nickel: experience of the British dermatologists (EPIDERM) and occupational physicians (OPRA) surveillance schemes, Shum KW	2003	✓			
Journal article—review, meta-analysis	1	Skin lesions and environmental exposures. An overview for the occupational health nurse, ATSDR	1996				✓

*See footnotes at end of table.

(Continued)

PROFESSIONAL AUDIENCE
Table 4B (Continued). Surveillance and clinical aspects

Resource type	ID	Title, author	Yr	Subtopics			
				B.1*	B.2	B.3	B.4
Journal article—review, meta-analysis (Continued)	27	Occupational issues of allergic contact dermatitis, Andersen KE	2003	✓			✓
	30	Occupational contact dermatitis, Antezana M	2003				✓
	38	Skin exposure to isocyanates: Reasons for concern, Bello D	2007	✓			
	42	In-use testing and interpretation of chemical-resistant glove performance, Boeniger MF	2002		✓		
	53	Strategies for prevention: Occupational contact dermatitis, Brown T	2004	✓	✓	✓	
	69	Occupational issues of irritant contact dermatitis, Chew AL	2003	✓	✓		✓
	73	Pesticide-related illness among migrant farm workers in the United States, Das R	2001	✓			
	74	Beryllium exposure: Dermal and immunological considerations, Day GA	2006	✓			
	76	What can we learn from epidemiological studies on irritant contact dermatitis? Diepgen TL	1995	✓		✓	
	77	Epidemiological studies on the prevention of occupational contact dermatitis, Diepgen TL	1996	✓			
	78	The epidemiology of occupational contact dermatitis, Diepgen TL	1999	✓			
	80	Occupational contact dermatitis I: Incidence and return to work pressures, Emmett EA	2002	✓	✓		
	81	Occupational contact dermatitis II: Risk assessment and prognosis, Emmett EA	2003	✓			✓

*See footnotes at end of table.

(Continued)

PROFESSIONAL AUDIENCE
Table 4B (Continued). Surveillance and clinical aspects

Resource type	ID	Title, author	Yr	B.1*	B.2	B.3	B.4
Journal article—review, meta-analysis (Continued)	101	Carbonless copy paper and workplace safety: A review, Graves CG	2000	✓			
	103	Occupation-related allergies in dentistry, Hamann CP	2005	✓			
	104	Textile dye allergic contact dermatitis prevalence, Hatch KL	2000	✓			
	122	A critique of assumptions about selecting chemical-resistant gloves: A case for workplace evaluation of glove efficacy, Klingner TD	2002	✓			
	128	Toxicity of methyl methacrylate in dentistry, Leggat PA	2003	✓			
	133	The epidemiology of occupational contact dermatitis, Lushniak BD	1995	✓	✓	✓	
	134	The importance of occupational skin diseases in the United States, Lushniak BD	2003	✓	✓		
	135	Occupational contact dermatitis, Lushniak BD	2004	✓	✓		
	146	Differences between the sexes with regard to work-related skin disease, Meding B	2000	✓			
	175	Clues to an accurate diagnosis of contact dermatitis, Rietschel RL	2004	✓			
	199	Nordic Occupational Skin Questionnaire-2002 (NOSQ-2002): A new tool for surveying occupational skin diseases and exposure, Susitaival P	2003			✓	
	203	Management of dermatitis in the rubber manufacturing industry, Toeppen-Sprigg B	1999	✓			

*See footnotes at end of table.

(Continued)

PROFESSIONAL AUDIENCE
Table 4B (Continued). Surveillance and clinical aspects

Resource type	ID	Title, author	Yr	B.1*	B.2	B.3	B.4
Journal article—review, meta-analysis (Continued)	223	Occupational contact dermatitis in the textile industry, Wigger-Alberti W	2003	✓			
	224	The dermal toxicity of cement, Winder C	2002	✓	✓		
Technical publication/report	129	Epidemiology of skin and respiratory diseases among hairdressers, FIOH	2001	✓		✓	
Web page	54	Bureau of Labor Statistics (BLS) industry illness and injury data [Web site], BLS	2005	✓	✓	✓	
	155	HSDB [Web site], NLM	2005	✓			
	220	Dermatitis: safety and health assessment and research for prevention (SHARP) [Home page] WADLI	2005	✓	✓	✓	
Web site	18	ATSDR [Home page], ATSDR	2005				✓
	57	California Division of Labor Statistics and Research [Home page], California Department of Industrial Relations [CA DIR]	2003	✓	✓		
	59	CPWR [Home page], CPWR	2006	✓	✓		
	82	Dermatological engineering [Web site], Enviroderm Services	2005			✓	✓
	87	United States Environmental Protection Agency [Home page], USEPA	2005	✓	✓	✓	
	105	Skin at work [Web site], HSE	2005	✓	✓		✓
	112	IW [Home page], ILO	2005	✓	✓		
	153	NIOSH [Home page], NIOSH	2005	✓	✓		

*See footnotes at end of table.

(Continued)

Chapter 4: Resources for the Professional Audience

PROFESSIONAL AUDIENCE
Table 4B (Continued). Surveillance and clinical aspects

Resource type	ID	Title, author	Yr	B.1*	B.2	B.3	B.4
Web site (Continued)	154	TOXNET—Databases on toxicology, hazardous chemicals, environmental health, and toxic releases [Home page], NLM	2005				✓
	164	Oregon Worker Illness and Injury Prevention Program (OWIIPP), Oregon Department of Human Services (ORDHS)	2005	✓	✓	✓	
	165	OR-OSHA [Home page], OR-OSHA	2006	✓			
	225	IPCS [Home page], WHO	2005				✓

*B.1=Surveillance Studies Reporting Incidences of Occupational Skin Exposures; B.2=Loss of Workdays and Impact on Productivity; B.3=Surveillance Study Protocols/Procedures for Gathering Data; B.4=Clinical Protocols for Recognition of Skin Exposure Health Effects

Topic 4C. Exposure Characterization

Exposure characterization, a component of exposure assessment, is the process of describing the conditions of a given occupational environment that may influence exposure. These conditions may include the source of the chemical; the magnitude, frequency, duration, and routes of the exposure to the chemical; the chemical and physical properties of the chemical; and work practices, or how work using the chemical is performed in a given working environment. The resources found in this table provide information associated with dermal exposure characterization to chemicals.

Subtopic C.1. Workplace Factors Associated with Harmful Skin Exposures

These resources contain information on workplace factors that influence the potential for skin exposure to chemicals in the workplace, including the tasks performed, industrial processes in which workers are engaged, chemicals used or produced, and occupations or job titles of individuals involved in that work.

Chapter 4: Resources for the Professional Audience

Subtopic C.2. Description of Factors Influencing Exposure Conditions

These resources provide quantitative descriptions of environmental factors that influence the potential for skin exposure, including exposure duration and frequency; exposure to chemical mixtures; the concentration of the chemical(s) to which the worker is exposed; the affected skin surface area that is exposed; and factors associated with chemical uptake through the skin.

Subtopic C.3. Checklists/Questionnaires to Quantify Skin Exposure Incidents

These resources contain either questionnaires, checklists or other tools or descriptions of tools that can be used to aid in the collection of quantitative exposure data by professionals, as well as in the reporting and characterization of skin exposures using quantitative data.

Subtopic C4. Direct Methods to Measure Exposures

These resources contain descriptions of methods that can be used to measure exposure. Exposure can be evaluated by measuring chemical contamination of the workplace environment surfaces on which skin contact occurs; measuring exposure to the skin by sampling the skin or skin surrogates, such as body suits, patches, tapes and strips; or by visualization techniques and performing biomonitoring.

Subtopic C.5. Exposure Modeling

These resources review exposure characterization methods using modeling based on predictive algorithms developed experimentally or using exposure estimates developed from exposure modeling.

PROFESSIONAL AUDIENCE
Table 4C. Exposure characterization

Resource type	ID	Title, author	Yr	C.1*	C.2	C.3	C.4	C.5
Book/monograph, chapter	4	Occupational skin exposure-absorption of chemical agents and assessment of exposures, Harris R	2000		✓			
	5	Dermal exposure modeling, Keil CB	2000		✓			✓

*See footnotes at end of table.

(Continued)

PROFESSIONAL AUDIENCE
Table 4C (Continued). Exposure characterization

Resource type	ID	Title, author	Yr	Subtopics				
				C.1*	C.2	C.3	C.4	C.5
Book/monograph, chapter (Continued)	28	Systemic toxicity from percutaneous absorption, Andersen KE	1999	✓	✓			
	90	Approaches for occupational dermal exposure assessment and management, Fehrenbacher MC	2003		✓		✓	✓
	158	Surface and dermal monitoring, Ness SA	2000		✓	✓	✓	✓
	171	Health risk assessment, Paustenbach D	1999		✓		✓	✓
	185	Dermal exposure assessments, Sahmel J	2006		✓		✓	✓
	221	Development of occupational skin disease, Weber LW	2003		✓			
Book/monograph, whole	17	Occupational skin disease, Adams RM	1999		✓		✓	✓
	137	Contact and occupational dermatology, Marks JG	2002				✓	
	157	Surface and dermal monitoring for toxic exposures, Ness SA	1994		✓	✓	✓	
	218	Health risk assessment: Dermal and inhalation exposure and absorption of toxicants (dermatology), Wang RGM	1993		✓			✓
	229	Dermatotoxicology, Zhai H	2004		✓			✓
Brochure, pamphlet	60	A safety and health practitioner's guide to skin protection, CPWR	2000	✓	✓	✓	✓	
Data file	200	Syracuse Research Corporation (SRC)—business areas: Environmental science [Home page], SRC	2006		✓		✓	
Guideline	47	Guideline for hand hygiene in healthcare Settings, CDC	2002					✓

*See footnotes at end of table.

(Continued)

PROFESSIONAL AUDIENCE
Table 4C (Continued). Exposure characterization

Resource type	ID	Title, author	Yr	Subtopics				
				C.1*	C.2	C.3	C.4	C.5
Guideline (Continued)	84	Organisation of Economic Co-operation and Development (OECD) series on testing and assessment, number 28: Guidance document for the conduct of skin absorption studies, OECD	2004		✓			
	161	Sampling for surface contamination, OSHA	2005				✓	
Journal article—primary	31	Effect of personal hygiene on blood lead levels of workers at a lead processing facility, Askin DP	1997	✓			✓	
	37	Slow curing of aliphatic polyisocyanate paints in automotive refinishing: A potential source for skin exposure, Bello D	2007		✓		✓	
	50	A dermal model for spray painters. Part I: Subjective exposure modeling of spray paint deposition, Brouwer DH	2001	✓	✓			✓
	65	Total body burden arising from a week's repeated dermal exposure to N,N-dimethylformamide (DMF)	2005		✓		✓	
	94	Modeling dermal exposure—an illustration for spray painting applications, Flynn MR	2006					✓
	96	An overview of human exposure modeling activities at the USEPA National Exposure Research Laboratory (NERL), Furtaw EJ Jr	2001					✓
	131	Skin exposure to aliphatic polyisocyanates in the auto body repair and refinishing industry: A qualitative assessment, Liu Y	2007		✓		✓	

*See footnotes at end of table.

(Continued)

PROFESSIONAL AUDIENCE
Table 4C (Continued). Exposure characterization

Resource type	ID	Title, author	Yr	Subtopics				
				C.1*	C.2	C.3	C.4	C.5
Journal article—primary (Continued)	142	Dermal exposure and urinary 1-hydroxypyrene among asphalt roofing workers, McClean MD	2007	✓	✓		✓	
	195	A dermal model for spray painters. Part II: Estimating the deposition and uptake of solvents, Semple S	2001	✓	✓			
	202	A real-time in-vivo method for studying the percutaneous absorption of volatile chemicals, Thrall KD	2000		✓		✓	✓
	214	DREAM: A method for semi-quantitative dermal exposure assessment, van-Wendel-de-Joode B	2003					✓
	215	Reliability of a semi-quantitative method for dermal exposure assessment (DREAM), van-Wendel-de-Joode B	2005				✓	✓
	227	Rapid method for determining dermal exposures to pesticides by use of tape stripping and FTIR spectroscopy: A pilot study, Wu CF	2007				✓	
Journal article—review, meta-analysis	1	Skin lesions and environmental exposures. An overview for the occupational health nurse, ATSDR	1996				✓	
	27	Occupational issues of allergic contact dermatitis, Andersen KE	2003				✓	
	30	Occupational contact dermatitis, Antezana M	2003				✓	
	38	Skin exposure to isocyanates: reasons for concern, Bello D	2007	✓	✓		✓	

*See footnotes at end of table.

(Continued)

PROFESSIONAL AUDIENCE
Table 4C (Continued). Exposure characterization

Resource type	ID	Title, author	Yr	C.1*	C.2	C.3	C.4	C.5
Journal article—review, meta-analysis (Continued)	39	Dermal route in systemic exposure, Benford DJ	1999		✓		✓	✓
	42	In-use testing and interpretation of chemical-resistant glove performance, Boeniger MF	2002				✓	✓
	44	Percutaneous absorption of organic solvents, Boman A	2000		✓			✓
	46	Proposal for the assessment of quantitative dermal exposure limits in occupational environments: Part 1. Development of a concept to derive a quantitative dermal occupational exposure limit, Bos PM	1998		✓			
	48	To what extent are biomonitoring data available in chemical risk assessment? Brondeau MT	1999		✓		✓	
	49	Hand wash and manual skin wipes, Brouwer DH	2000				✓	
	51	Concepts of skin protection: Considerations for the evaluation and terminology of the performance of skin protective equipment, Brouwer DH	2005		✓			✓
	56	Suction methods for assessing contamination on surfaces, Byrne MA	2000				✓	✓
	67	Use of qualitative and quantitative fluorescence techniques to assess dermal exposure, Cherrie JW	2000				✓	
	69	Occupational issues of irritant contact dermatitis, Chew AL	2003				✓	

*See footnotes at end of table.

(Continued)

PROFESSIONAL AUDIENCE
Table 4C (Continued). Exposure characterization

Resource type	ID	Title, author	Yr	Subtopics				
				C.1*	C.2	C.3	C.4	C.5
Journal article—review, meta-analysis (Continued)	76	What can we learn from epidemiological studies on irritant contact dermatitis? Diepgen TL	1995	✓				
	81	Occupational contact dermatitis II: risk assessment and prognosis, Emmett EA	2003		✓		✓	
	91	Dermal exposure assessment techniques, Fenske RA	1993	✓			✓	✓
	93	Modelling skin permeability in risk assessment—the future, Fitzpatrick D	2004	✓			✓	
	97	Developing COSHH Essentials: dermal exposure, personal protective equipment and first aid, Garrod AN	2003	✓			✓	
	104	Textile dye allergic contact dermatitis prevalence, Hatch KL	2000				✓	
	107	Factors determining percutaneous metal absorption, Hostynek JJ	2003	✓				
	115	Dermal absorption of benzene: implications for work practices and regulations, Kalnas J	2000	✓			✓	
	120	Improved estimation of dermal pesticide dose to agricultural workers upon reentry, Kissel J	2000	✓			✓	
	121	Factors affecting soil adherence to skin in hand-press trials, Kissel JC	1996	✓				
	124	Occupational contact dermatitis. Recognition and management, Koch P	2001	✓				

*See footnotes at end of table.

(Continued)

PROFESSIONAL AUDIENCE
Table 4C (Continued). Exposure characterization

Resource type	ID	Title, author	Yr	C.1*	C.2	C.3	C.4	C.5
Journal article—review, meta-analysis (Continued)	126	Temporal, personal and spatial variability in dermal exposure, Kromhout H	2001	✓	✓			✓
	130	Techniques for estimating the percutaneous absorption of chemicals due to occupational and environmental exposure, Leung H-W	1994		✓		✓	✓
	132	Percutaneous absorption of arsenic from environmental media, Lowney YW	2005		✓			
	135	Occupational contact dermatitis, Lushniak BD	2004	✓				
	138	Harmonization of future needs for dermal exposure assessment and modeling: A workshop report, Marquart H	2001		✓		✓	✓
	139	Determinants of dermal exposure relevant for exposure modeling in regulatory risk assessment, Marquart J	2003	✓	✓			✓
	141	Dermal Measurement and Wipe Sampling Methods: A Review, McArthur B	1992				✓	
	143	Assessment of dermal absorption and penetration of components of a fuel mixture (JP-8), McDougal JN	2002		✓			
	144	Methods for assessing risks of dermal exposures in the workplace, McDougal JN	2002		✓			✓
	148	Quantitative structure-permeability relationships (QSPRs) for percutaneous absorption, Moss GP	2002		✓			
	160	New approaches for assessment of occupational exposure to metals using on-site measurements, Nygren O	2002				✓	

*See footnotes at end of table.

(Continued)

PROFESSIONAL AUDIENCE
Table 4C (Continued). Exposure characterization

Resource type	ID	Title, author	Yr	C.1*	C.2	C.3	C.4	C.5
Journal article—review, meta-analysis (Continued)	163	A toolkit for dermal risk assessment and management: an overview, Oppl R	2003				✓	
	172	Assessment of dermal exposure—empirical models and indicative distributions, Phillips AM	2001				✓	
	173	Assessing dermal absorption, Poet TS	2000		✓			
	175	Clues to an accurate diagnosis of contact dermatitis, Rietschel RL	2004				✓	
	177	Quantitating absorption of complex chemical mixtures, Riviere JE	2004		✓			
	182	Conservatism in pesticide exposure assessment, Ross JH	2000		✓			
	183	Solvents and the skin, Rowse DH	2004		✓			
	184	A distributed parameter physiologically-based pharmacokinetic model for dermal and inhalation exposure to volatile organic compounds, Roy A	1996		✓			✓
	187	Percutaneous penetration studies for risk assessment, Sartorelli P	2000		✓			
	188	Dermal exposure assessment in occupational medicine, Sartorelli P	2002		✓			
	190	Conceptual model for assessment of dermal exposure, Schneider T	1999		✓		✓	✓
	191	Dermal exposure assessment, Schneider T	2000				✓	✓
	194	Dermal exposure to chemicals in the workplace: Just how important is skin absorption? Semple S	2004		✓		✓	✓

*See footnotes at end of table.

(Continued)

PROFESSIONAL AUDIENCE
Table 4C (Continued). Exposure characterization

Resource type	ID	Title, author	Yr	Subtopics				
				C.1*	C.2	C.3	C.4	C.5
Journal article— review, meta- analysis (Continued)	197	From xenobiotic chemistry and metabolism to better prediction and risk assessment of skin allergy, Smith Pease CK	2003		✓			
	198	Use of patches and whole body sampling for the assessment of dermal exposure, Soutar A	2000				✓	✓
	204	Prediction of irritancy in the human skin irritancy model and occupational setting, Tupker RA	2003		✓		✓	
	210	From dermal exposure to internal dose, van de Sandt J	2008				✓	✓
	211	Assessment of dermal exposure to chemicals, van Hemmen JJ	1995		✓		✓	
	213	Review of skin permeation hazard of bitumen fumes, van Rooij JG	2008		✓			
	216	Dermal exposure assessment in occupational epidemiologic research, Vermeulen R	2002		✓			✓
	219	Deriving default dermal exposure values for use in a risk assessment toolkit for small and medium-sized enterprises, Warren N	2003					✓
	222	Understanding percutaneous absorption for occupational health and safety, Wester RC	2000		✓			✓
	228	Toxic effects of chemical mixtures, Zeliger HI	2003		✓			

*See footnotes at end of table.

(Continued)

PROFESSIONAL AUDIENCE
Table 4C (Continued). Exposure characterization

Resource type	ID	Title, author	Yr	Subtopics				
				C.1*	C.2	C.3	C.4	C.5
Other—commentary	41	Exposure and absorption of hazardous materials through the skin, Boeniger MF	2000		✓			
Technical publication/report	10	Risk Assessment Guidance for Superfund (RAGS), Volume I: Human Health Evaluation Manual (Part E, Supplemental Guidance for Dermal Risk Assessment)	2001		✓			✓
	36	Dermal absorption of cutting fluid mixtures, Baynes RE	2003		✓			
	66	Occupational dermal exposure assessment: A review of methodologies and field data-final report, Chen CK	1996				✓	✓
	86	Summary report for the workshop on issues associated with dermal exposure and uptake, USEPA, Bethesda, MD, December 10–11, 1998, United States	2000		✓		✓	✓
	89	Skin and respiratory sensitizers: reference chemicals data bank, European Centre for Ecotoxicology and Toxicology of Chemicals (ECETOC)	1999					✓
	102	Prediction and assessment of dermal exposure, Guy R	1998		✓			✓
	108	Dermal and non-dietary ingestion exposure workshop: NERL Human Exposure Research Program, Hubal EC	1998		✓		✓	✓
	114	CEFIC workshop on methods to determine dermal permeation for human risk assessment, IOM	2004		✓			✓
	129	Epidemiology of skin and respiratory diseases among hairdressers, FIOH	2001	✓			✓	

*See footnotes at end of table.

(Continued)

Chapter 4: Resources for the Professional Audience

PROFESSIONAL AUDIENCE
Table 4C (Continued). Exposure characterization

Resource type	ID	Title, author	Yr	C.1*	C.2	C.3	C.4	C.5
Technical publication/report (Continued)	176	Percutaneous absorption of chemical mixtures relevant to the Gulf War, Riviere JE	2002		✓			
	178	Quantitating the percutaneous absorption of mechanistically defined chemical mixtures final report 15 Nov 1997–14 Nov 2000, Riviere JE	2001		✓			
	179	Quantitating the percutaneous absorption of mechanistically defined chemical mixtures final report 15 Dec 2000–14 Dec 2003, Riviere JE	2004	✓	✓			
	226	IPCS: Environmental health criteria document on dermal absorption [Draft], WHO	2005		✓			
Web page	19	Toxicological profile information sheet [Web site], ATSDR	2005				✓	
	85	Harmonized test guidelines [Web site], USEPA	1998				✓	
	155	HSDB [Web site], NLM	2005	✓				
	167	OECD's database on chemical risk assessment models, OECD	2006					✓
	209	USEPA, Office of Pollution Prevention and Toxics (OPPT): Exposure assessment tools and models [Home page], USEPA	2005					✓
	220	Dermatitis: Safety and health assessment and research for prevention (SHARP) [Home page] WADLI	2005	✓				
Web site	16	CPI [Web site], ACC	2006	✓				

*See footnotes at end of table.

(Continued)

PROFESSIONAL AUDIENCE
Table 4C (Continued). Exposure characterization

Resource type	ID	Title, author	Yr	Subtopics				
				C.1*	C.2	C.3	C.4	C.5
Web site (Continued)	32	American Society for Testing and Materials (ASTM) International [Home page], ASTM International	2006	✓			✓	
	71	The pioneer in the reduction of dermal exposure [Colormetric Laboratories, Inc. Home page], Colormetric Laboratories, Inc.	2005				✓	
	82	Dermatological engineering [Web site], Enviroderm Services	2005	✓	✓		✓	
	87	USEPA [Home page], USEPA	2005		✓		✓	✓
	105	Skin at work [Web site], HSE	2005	✓	✓		✓	
	112	ILO [Home page], ILO	2005	✓	✓			
	153	NIOSH [Home page], NIOSH	2005	✓	✓		✓	
	154	TOXNET—Databases on toxicology, hazardous chemicals, environmental health, and toxic releases [Home page], NLM	2005	✓				
	162	OSHA [Home page], OSHA	2005	✓	✓		✓	
	201	Wil ten Berge model for dermal absorption [Home page], ten Berge W	2004					✓

*C.1=Workplace Factors Associated with Harmful Skin Exposures; C.2=Description of Factors Influencing Exposure Conditions; C.3=Checklists/Questionnaires to Quantify Skin Exposure Incidents; C.4=Methods to Measure Exposures; C.5=Exposure Modeling

Topic 4D. Hazard Identification from Toxicological Studies or Modeling

Hazard identification, another component of exposure risk assessment, is the process of establishing the existence of a hazard through field observations and laboratory analysis of the exposures and adverse health effects. As part of this process, the nature of the hazard is determined, such as whether the chemicals causes skin irritation, skin corrosion, sensitization or some systemic toxic effect, and the dose-response relationship under the conditions of exposure is determined. Resources found in the following tables review information associated with hazard identification based on toxicological studies or modeling efforts.

Subtopic D.1. Potential Health Effects Resulting from Specific Chemicals

These resources review or summarize information on potential health effects resulting from skin exposure to chemicals. Health effects can include localized skin irritation and corrosion, including irritant contact dermatitis; sensitization of the skin, including allergic contact dermatitis, or sensitization of the respiratory tract as a result of skin exposure; the potential contribution of skin exposure and resulting dermal absorption to systemic toxicity; or the contribution of skin exposure to overall or total exposure to a chemical(s).

Subtopic D.1.a. Irritant Contact Dermatitis

These resources address chemically induced irritant contact dermatitis. This form of dermatitis is caused by direct exposure of the skin to a chemical or other irritating substance. Irritation in the form of inflammation usually occurs either immediately or within a short period of time.

Subtopic D.1.b. Allergic Contact Dermatitis/Sensitization

Resources found here include information on allergic contact dermatitis, which is an immunologically mediated reaction of the skin caused by direct contact of the skin of a sensitized individual to a chemical that is an allergen. Inflammation also occurs, but only in sensitized individuals.

Subtopic D.1.c. Systemic Toxicity

Resources listed under this subtopic include information on dermal exposures that can cause systemic toxicity. Unlike irritant and allergic contact dermatitis, which cause localized toxic effects, systemic toxicity refers to chemicals absorbed through the skin which are then transported to other sites in the body where their toxic effects occur. It is possible for a chemical to produce both a local and systemic adverse health effect.

Subtopic D.1.d. Other Health Effects

These resources contain information on other health effects that can be caused by dermal exposure and include conditions such as urticaria, which is immediately hypersensitivity; foreign body dermatitis, caused by foreign compounds such as fiber glass, silica, and asbestos penetrating the skin; pigmentation changes of the skin; cancer; and other health effects.

Subtopic D.2. Summaries of Health Effects and Dose-Response Relationships

These resources contain sources of data or actual databases that provide summaries and discuss the significance of health effects that result from skin exposures to chemicals in the workplace.

Subtopic D.3. Characterization Protocols

These resources contain protocols or guidelines for use in chemical hazard characterizations. Protocols and guidelines cited include those for various kinds of toxicological studies. The responses characterized in these protocols and guidelines include corrosivity, irritation potential, sensitization potential, and potential to cause systemic effects. Also included are resources that provide protocols and guidelines for measuring skin permeation rates and reservoir effects, as well as protocols for developing and validating qualitative and quantitative structure activity relationships QSAR for application in hazard identification and for use in validating QSAR as a screening tool to identify skin hazards in high priority chemicals.

PROFESSIONAL AUDIENCE

Table 4D(I). Hazard identification from toxicological studies or modeling

Resource type	ID	Title, author	Yr	D.1a*	D.1b	D.1c	D.1d
Book/monograph, chapter	4	Occupational skin exposure—absorption of chemical agents and assessment of exposures, Harris R	2000	✓	✓		
	28	Systemic toxicity from percutaneous absorption, Andersen KE	1999		✓	✓	
	40	CPC and the skin: Practical considerations, Boeniger M	2002	✓			
	221	Development of occupational skin disease, Weber LW	2003	✓	✓		

*See footnotes at end of table.

(Continued)

PROFESSIONAL AUDIENCE
Table 4D(I) (Continued). Hazard identification from toxicological studies or modeling

				D Subtopics			
Resource type	ID	Title, author	Yr	D.1a*	D.1b	D.1c	D.1d
Book/monograph, whole	17	Occupational skin disease, Adams RM	1999	✓	✓	✓	✓
	117	Handbook of occupational dermatology, Kanerva L	2000	✓	✓	✓	
	137	Contact and occupational dermatology, Marks JG	2002	✓	✓		
	218	Health risk assessment: Dermal and inhalation exposure and absorption of toxicants (dermatology), Wang RGM	1993				✓
	229	Dermatotoxicology, Zhai H	2004	✓	✓	✓	✓
Brochure, pamphlet	60	A safety and health practitioner's guide to skin protection, CPWR	2000	✓	✓		✓
Data file	200	SRC—business areas: Environmental science [Home page], SRC	2006	✓	✓		
Guideline	23	Documentation of the threshold limit values for chemical substances, ACGIH	2001	✓	✓	✓	✓
Journal article—review, meta-analysis	1	Skin lesions and environmental exposures. An overview for the occupational health nurse, ATSDR	1996	✓	✓	✓	✓
	26	Cleansing without compromise: The impact of cleansers on the skin barrier and the technology of mild cleansing, Ananthapadmanabhan KP	2004	✓			
	30	Occupational contact dermatitis, Antezana M	2003	✓	✓		
	34	Classification criteria for skin-sensitizing chemicals: a commentary, Basketter DA	1999	✓	✓	✓	

*See footnotes at end of table.

(Continued)

PROFESSIONAL AUDIENCE
Table 4D(I) (Continued). Hazard identification from toxicological studies or modeling

Resource type	ID	Title, author	Yr	D.1a*	D.1b	D.1c	D.1d
Journal article—review, meta-analysis (Continued)	38	Skin exposure to isocyanates: Reasons for concern, Bello D	2007		✓		✓
	46	Proposal for the assessment of quantitative dermal exposure limits in occupational environments: Part 1. Development of a concept to derive a quantitative dermal occupational exposure limit, Bos PM	1998	✓	✓		
	69	Occupational issues of irritant contact dermatitis, Chew AL	2003	✓			
	74	Beryllium exposure: Dermal and immunological considerations, Day GA	2006		✓	✓	
	77	Epidemiological studies on the prevention of occupational contact dermatitis, Diepgen TL	1996		✓		
	78	The epidemiology of occupational contact dermatitis, Diepgen TL	1999	✓	✓		
	79	Skin-conditioning products in occupational dermatology, Elsner P	2003	✓			
	81	Occupational contact dermatitis II: Risk assessment and prognosis, Emmett EA	2003	✓	✓		
	99	A chemical dataset for evaluation of alternative approaches to skin-sensitization testing, Gerberick GF	2004		✓		
	101	CCP and workplace safety: A review, Graves CG	2000	✓			
	103	Occupation-related allergies in dentistry, Hamann CP	2005	✓	✓		

*See footnotes at end of table.

(Continued)

PROFESSIONAL AUDIENCE
Table 4D(I) (Continued). Hazard identification from toxicological studies or modeling

Resource type	ID	Title, author	Yr	D Subtopics			
				D.1a*	D.1b	D.1c	D.1d
Journal article—review, meta-analysis (Continued)	116	Dermatological aspects of a successful introduction and continuation of alcohol-based hand rubs for hygienic hand disinfection, Kampf G	2003	✓			
	118	The role of the skin in the development of chemical respiratory hypersensitivity, Kimber I	1996		✓	✓	✓
	119	Alternative approaches to the identification and characterization of chemical allergens, Kimber I	2001		✓		
	122	A critique of assumptions about selecting chemical-resistant gloves: A case for workplace evaluation of glove efficacy, Klingner TD	2002	✓	✓	✓	
	124	Occupational contact dermatitis. Recognition and management, Koch P	2001	✓			
	125	Occupational skin-protection products—a review, Kresken J	2003	✓	✓		
	127	Effectiveness of skin protection creams as a preventive measure in occupational dermatitis: A critical update according to criteria of evidence-based medicine, Kutting B	2003	✓			
	128	Toxicity of methyl methacrylate in dentistry, Leggat PA	2003			✓	
	133	The epidemiology of occupational contact dermatitis, Lushniak BD	1995	✓			
	135	Occupational contact dermatitis, Lushniak BD	2004	✓	✓		✓

*See footnotes at end of table.

(Continued)

PROFESSIONAL AUDIENCE
Table 4D(I) (Continued). Hazard identification from toxicological studies or modeling

Resource type	ID	Title, author	Yr	D Subtopics			
				D.1a*	D.1b	D.1c	D.1d
Journal article—review, meta-analysis (Continued)	140	Dermal toxicity due to industrial chemicals, Mathur AK	2002	✓	✓	✓	✓
	146	Differences between the sexes with regard to work-related skin disease, Meding B	2000	✓	✓		
	175	Clues to an accurate diagnosis of contact dermatitis, Rietschel RL	2004	✓	✓		
	180	Quantitative structure-activity relationships for predicting skin and respiratory sensitization, Rodford R	2003		✓		
	183	Solvents and the skin, Rowse DH	2004	✓	✓	✓	✓
	189	Chemical substances and contact allergy—244 substances ranked according to allergenic potency, Schlede E	2003		✓		
	192	When should a substance be designated as sensitizing for the skin ('Sh') or for the airways ('Sa')?, Schnuch A	2002		✓		
	197	From xenobiotic chemistry and metabolism to better prediction and risk assessment of skin allergy, Smith Pease CK	2003		✓		
	213	Review of skin permeation hazard of bitumen fumes, van Rooij JG	2008				✓
	223	Occupational contact dermatitis in the textile industry, Wigger-Alberti W	2003	✓	✓		
	224	The dermal toxicity of cement, Winder C	2002	✓	✓		✓
Technical publication/report	36	Dermal absorption of cutting fluid mixtures, Baynes RE	2003	✓			

*See footnotes at end of table.

(Continued)

PROFESSIONAL AUDIENCE
Table 4D(I) (Continued). Hazard identification from toxicological studies or modeling

Resource type	ID	Title, author	Yr	D Subtopics			
				D.1a*	D.1b	D.1c	D.1d
Technical publication/report (Continued)	89	Skin and respiratory sensitizers: Reference chemicals data bank, European Centre for Ecotoxicology and Toxicology of Chemicals	1999	✓	✓		
	129	Epidemiology of skin and respiratory diseases among hairdressers, FIOH	2001	✓	✓		✓
	179	Quantitating the percutaneous absorption of mechanistically defined chemical mixtures, 15 Dec 2000–14 Dec 2003, Riviere JE	2004			✓	
Web page	19	Toxicological profile information sheet [Web page], ATSDR	2005	✓	✓	✓	✓
	155	HSDB [Web page], NLM	2005	✓	✓	✓	✓
Web site	16	CPI [Web site], ACC	2006		✓		
	18	ATSDR [Home page], ATSDR	2005	✓	✓	✓	✓
	22	AAFP [Home page], AAFP	2005	✓	✓		
	32	ASTM International [Home page], ASTM International	2006		✓		
	82	Dermatological engineering [Web page], Enviroderm Services	2005	✓	✓	✓	
	105	Skin at work [Web page], HSE	2005	✓	✓		
	110	Interagency Coordinating Committee on the Validation of Alternative Methods (ICCVAM) [Home page], ICCVAM	2005		✓		
	112	ILO [Home page], ILO	2005	✓	✓	✓	
	153	NIOSH [Home page], NIOSH	2005	✓	✓	✓	

*See footnotes at end of table.

(Continued)

PROFESSIONAL AUDIENCE
Table 4D(I) (Continued). Hazard identification from toxicological studies or modeling

Resource type	ID	Title, author	Yr	D.1a*	D.1b	D.1c	D.1d
Web site (Continued)	154	TOXNET—Databases on toxicology, hazardous chemicals, environmental health, and toxic releases [Home page], NLM	2005	✓	✓	✓	✓
	206	EXTOXNET [Web site], Ohio State University	2006	✓	✓	✓	
	225	IPCS [Home page], WHO	2005	✓	✓	✓	✓

*D1=Potential Health Effects Resulting from Specific Chemicals; D1a=Irritant Contact Dermatitis; D1b=Allergic Contact Dermatitis/Sensitization; D1c=Systemic Toxicity; D1d=Other Health Effects

PROFESSIONAL AUDIENCE
Table 4D(II). Hazard identification from toxicological studies or modeling

Resource type	ID	Title, author	Yr	D.2*	D.3
Book/monograph, chapter	171	Health risk assessment, Paustenbach D	1999	✓	✓
Book/monograph, whole	17	Occupational skin disease, Adams RM	1999		✓
	137	Contact and occupational dermatology, Marks JG	2002		✓
	218	Health risk assessment: dermal and inhalation exposure and absorption of toxicants (dermatology), Wang RGM	1993	✓	
Data file	200	SRC—business areas: environmental science [Home page], SRC	2006		✓
Guideline	2	Health effects test guidelines: OPPTS 870.2500, acute dermal irritation, OPPTS	1998		✓

*See footnotes at end of table.

(Continued)

PROFESSIONAL AUDIENCE
Table 4D(II) (Continued). Hazard identification from toxicological studies or modeling

Resource type	ID	Title, author	Yr	Subtopics D.2*	D.3
Guideline (Continued)	23	Documentation of the threshold limit values for chemical substances, ACGIH	2001	✓	
	84	OECD series on testing and assessment, number 28: guidance document for the conduct of skin absorption studies, OECD	2004		✓
Journal article—primary	65	Total body burden arising from a week's repeated dermal exposure to N,N-dimethylformamide (DMF), Chang H	2005	✓	
	131	Skin exposure to aliphatic polyisocyanates in the auto body repair and refinishing industry: A qualitative assessment, Liu Y	2007		✓
	142	Dermal exposure and urinary 1-hydroxypyrene among asphalt roofing workers, McClean MD	2007		✓
	202	A real-time in-vivo method for studying the percutaneous absorption of volatile chemicals, Thrall KD	2000	✓	✓
Journal article—review, meta-analysis	27	Occupational issues of allergic contact dermatitis, Andersen KE	2003	✓	
	35	Factors affecting thresholds in allergic contact dermatitis: safety and regulatory considerations, Basketter DA	2002	✓	
	39	Dermal route in systemic exposure, Benford DJ	1999		✓
	44	Percutaneous absorption of organic solvents, Boman A	2000	✓	
	46	Proposal for the assessment of quantitative dermal exposure limits in occupational environments: Part 1. Development of a concept to derive a quantitative dermal occupational exposure limit, Bos PM	1998		✓
	77	Epidemiological studies on the prevention of occupational contact dermatitis, Diepgen TL	1996	✓	

*See footnotes at end of table.

(Continued)

PROFESSIONAL AUDIENCE
Table 4D(II) (Continued). Hazard identification from toxicological studies or modeling

				Subtopics	
Resource type	ID	Title, author	Yr	D.2*	D.3
Journal article—review, meta-analysis (Continued)	78	The epidemiology of occupational contact dermatitis, Diepgen TL	1999		✓
	81	Occupational contact dermatitis II: Risk assessment and prognosis, Emmett EA	2003	✓	
	93	Modeling skin permeability in risk assessment—the future, Fitzpatrick D	2004		✓
	99	A chemical dataset for evaluation of alternative approaches to skin-sensitization testing, Gerberick GF	2004		✓
	101	CCP and workplace safety: A review, Graves CG	2000		✓
	119	Alternative approaches to the identification and characterization of chemical allergens, Kimber I	2001		✓
	130	Techniques for estimating the percutaneous absorption of chemicals due to occupational and environmental exposure, Leung H-W	1994		✓
	148	Quantitative structure-permeability relationships (QSPRs) for percutaneous absorption, Moss GP	2002		✓
	170	Quantitative structure-activity relationships for predicting skin and eye irritation, Patlewicz G	2003		✓
	175	Clues to an accurate diagnosis of contact dermatitis, Rietschel RL	2004	✓	
	177	Quantitating absorption of complex chemical mixtures, Riviere JE	2004		✓
	180	Quantitative structure-activity relationships for predicting skin and respiratory sensitization, Rodford R	2003		✓
	183	Solvents and the skin, Rowse DH	2004	✓	✓

*See footnotes at end of table.

(Continued)

PROFESSIONAL AUDIENCE
Table 4D(II) (Continued). Hazard identification from toxicological studies or modeling

Resource type	ID	Title, author	Yr	Subtopics D.2*	D.3
Journal article—review, meta-analysis (Continued)	187	Percutaneous penetration studies for risk assessment, Sartorelli P	2000		✓
	189	Chemical substances and contact allergy—244 substances ranked according to allergenic potency, Schlede E	2003		✓
	197	From xenobiotic chemistry and metabolism to better prediction and risk assessment of skin allergy, Smith Pease CK	2003	✓	✓
	210	From dermal exposure to internal dose, van de Sandt J, Dellarco M	2008		✓
	217	Quantitative structure-activity relationships for predicting percutaneous absorption rates, Walker JD	2003		✓
	228	Toxic effects of chemical mixtures, Zeliger HI	2003		✓
Technical publication/report	36	Dermal absorption of cutting fluid mixtures, Baynes RE	2003		✓
	66	Occupational dermal exposure assessment: a review of methodologies and field data: Final report, Chen CK	1996	✓	
	114	CEFIC Workshop on methods to determine dermal permeation for human risk assessment, IOM	2004		✓
	176	Percutaneous absorption of chemical mixtures relevant to the Gulf War, Riviere JE	2002		✓
	178	Quantitating the percutaneous absorption of mechanistically defined chemical mixtures: 15 Nov 1997–14 Nov 2000, Riviere JE	2001		✓
	179	Quantitating the percutaneous absorption of mechanistically defined chemical mixtures: 15 Dec 2000–14 Dec 2003, Riviere JE, Monteiro-Riviere NA	2004	✓	

*See footnotes at end of table.

(Continued)

PROFESSIONAL AUDIENCE
Table 4D(II) (Continued). Hazard identification from toxicological studies or modeling

				Subtopics	
Resource type	ID	Title, author	Yr	D.2*	D.3
Technical publication/ report (Continued)	226	The International Programme on Chemical Safety: Environmental health criteria document on dermal absorption [Draft], WHO	2005		✓
Web page	19	Toxicological profile information sheet [Home page], ATSDR	2005	✓	
	85	Harmonized test guidelines [Web page], USEPA	1998		✓
	166	Guidelines for the testing of chemicals [Web page], OECD	2005		✓
	167	OECD's database on chemical risk assessment models [Web page], OECD	2006		✓
Web site	18	ATSDR [Home page], ATSDR	2005	✓	
	87	USEPA [Home page], USEPA	2005		✓
	105	Skin at work [Web site], HSE	2005	✓	✓
	110	ICCVAM [Home page], ICCVAM	2005		✓
	154	TOXNET—Databases on toxicology, hazardous chemicals, environmental health, and toxic releases [Home page], NLM	2005	✓	
	201	Wil ten Berge model for dermal absorption [Home page], ten Berge W	2004		✓
	206	EXTOXNET—The Extension Toxicology Network [Web site], Ohio State University	2006	✓	
	225	IPCS [Home page], WHO	2005	✓	✓

*D2=Summaries of Health Effects and Dose-Response Relationships; D3=Characterization Protocols

Topic 4E. Risk Assessment

Risk assessment is the measurement or estimate of the chances of a given level of exposure to a chemical to cause harm. With respect to skin exposures, risk assessments are performed by workplace health and safety professionals to provide the employer with an estimate of the likelihood of an illness or injury to result from skin exposure to a chemical hazard. Risk assessment is essential for setting occupational safety and health priorities and for demonstrating health impairment when promulgating occupational standards. These resources contain information associated with skin exposure risk assessments.

Subtopic E.1. Guidelines for Risk Assessment or Analysis

The resources indicated in this column provide descriptions of risk assessments and guildlines for analyses to determine if skin exposure to a chemical likely to cause a given effect, either a localized health effect or a systemic health effect.

Subtopic E.2. Example of Risk Assessments

The resources identified in this column provide examples of risk assessments that have been conducted to evaluate hazards associated with a specific chemical or chemical mixture in addition to models used to evaluate risk of dermal exposures.

PROFESSIONAL AUDIENCE
Table 4E. Risk assessment

Resource type	ID	Title, author	Yr	Subtopics E.1*	E.2
Book/monograph, chapter	90	Approaches for occupational dermal exposure assessment and management, Fehrenbacher MC	2003	✓	
	171	Health risk assessment, Paustenbach D	1999	✓	✓
	185	Dermal exposure assessments, Sahmel J	2006	✓	✓
Book/monograph, whole	229	Dermatotoxicology, Zhai H	2004	✓	
Brochure, pamphlet	60	A safety and health practitioner's guide to skin protection, CPWR	2000	✓	

*See footnotes at end of table.

(Continued)

PROFESSIONAL AUDIENCE
Table 4E (Continued). Risk assessment

Resource type	ID	Title, author	Yr	Subtopics E.1*	E.2
Journal article—primary	33	A structured strategy for assessing chemical risks, suitable for small and medium-sized enterprises, Balsat A	2003	✓	✓
	96	An overview of human exposure modeling activities at the USEPA National Exposure Research Laboratory, Furtaw EJ Jr	2001		✓
	131	Skin exposure to aliphatic polyisocyanates in the auto body repair and refinishing industry: A qualitative assessment, Liu Y	2007		
	168	Risk assessment and exposure control in an occupational setting, Packham CL	1996	✓	✓
Journal article—review, meta-analysis	81	Occupational contact dermatitis II: Risk assessment and prognosis, Emmett EA	2003	✓	
	100	Classification of dermal exposure modifiers and assignment of values for a risk assessment toolkit, Goede HA	2003	✓	
	106	Misinterpretation and misuse of exposure limits, Hewett P	2001	✓	
	115	Dermal absorption of benzene: Implications for work practices and regulations, Kalnas J	2000	✓	✓
	126	Temporal, personal and spatial variability in dermal exposure, Kromhout H	2001		✓
	139	Determinants of dermal exposure relevant for exposure modelling in regulatory risk assessment, Marquart J	2003	✓	
	143	Assessment of dermal absorption and penetration of components of a fuel mixture (JP-8), McDougal JN	2002		✓
	163	A toolkit for dermal risk assessment and management: an overview, Oppl R	2003	✓	
	172	Assessment of dermal exposure—empirical models and indicative distributions, Phillips AM	2001		✓

*See footnotes at end of table.

(Continued)

PROFESSIONAL AUDIENCE
Table 4E (Continued). Risk assessment

Resource type	ID	Title, author	Yr	Subtopics E.1*	E.2
Journal article—review, meta-analysis (Continued)	177	Quantitating absorption of complex chemical mixtures, Riviere JE	2004		✓
	182	Conservatism in pesticide exposure assessment, Ross JH	2000	✓	
	187	Percutaneous penetration studies for risk assessment, Sartorelli P	2000	✓	
	193	A toolkit for dermal risk assessment: Toxicological approach for hazard characterization, Schuhmacher-Wolz U	2003	✓	
	210	From dermal exposure to internal dose, van de Sandt J	2008		✓
	212	RISKOFDERM: Risk assessment of occupational dermal exposure to chemicals. An introduction to a series of papers on the development of a toolkit, van Hemmen JJ	2003	✓	
	219	Deriving default dermal exposure values for use in a risk assessment toolkit for small and medium-sized enterprises, Warren N	2003	✓	
Technical publication/report	114	CEFIC Workshop on methods to determine dermal permeation for human risk assessment, IOM	2004		✓
Web page	209	USEPA, OPPT: Exposure assessment tools and models, USEPA	2005	✓	✓
Web site	71	The pioneer in the reduction of dermal exposure [Colormetric Laboratories, Inc. Home page], Colormetric Laboratories, Inc.	2005	✓	
	82	Dermatological engineering [Web page], Enviroderm Services	2005	✓	✓
	87	USEPA [Home page], USEPA	2005	✓	✓
	105	Skin at work [Web page], HSE	2005	✓	✓

*See footnotes at end of table.

(Continued)

PROFESSIONAL AUDIENCE
Table 4E (Continued). Risk assessment

Resource type	ID	Title, author	Yr	Subtopics E.1*	E.2
Web site (Continued)	109	Human Exposure Research Organizations Exchange [HEROX] [Home page], HEROX	2005	✓	
	162	OSHA [Home page], OSHA	2005	✓	

*E.1=Guidelines for Risk Assessment or Analysis; E.2=Example of Risk Assessments

Topic 4F. Risk Management

Risk management completes the process of addressing potential hazards in the workplace, using information from the process of hazard identification and characterization and then taking steps to eliminate or control the hazard by various techniques. Control techniques can include changes in the process to (1) reduce or eliminate the hazard, (2) replace a more harmful chemical with a less harmful one, (3) isolate a process to minimize worker contact with the hazards, (4) modify the source to achieve less hazardous conditions/exposures, (5) change work practices to make the task less hazardous, (6) put in place administrative controls such as worker rotation, (7) train and monitor to lower each workers exposure to the hazards, or (8) use personal protective equipment to lower the exposure to the hazard. Risk management also involves evaluating the effectiveness of controls taken. The resources contain information associated with risk management.

Subtopic F. 1. Exposure Control Strategies

These resources review strategies that can be used to control skin exposures to chemicals. The following categories of control strategies are discussed: substitution; engineering controls; work practices and administrative controls; personal protective equipment; and the effectiveness of skin management programs using barrier creams, moisturizers, cleansers, and rubs.

Subtopic F.2. Protocols for Risk Management

These resources contain protocols for risk management programs. They may include protocols for the development of exposure reduction goals, protocols for the development of approaches to achieve exposure reduction goals, and evaluation tools to demonstrate program or intervention effectiveness.

Chapter 4: Resources for the Professional Audience

PROFESSIONAL AUDIENCE
Table 4F. Risk management

Resource type	ID	Title, author	Yr	Subtopics F.1*	F.2
Book/monograph, chapter	4	Occupational skin exposure—absorption of chemical agents and assessment of exposures, Harris R	2000	✓	
	40	CPC and the skin: Practical considerations, Boeniger M	2002	✓	
Book/monograph, chapter	90	Approaches for occupational dermal exposure assessment and management, Fehrenbacher MC	2003	✓	
	136	Personal protective clothing, Mansdorf SZ	2003	✓	
Book/monograph, whole	17	Occupational skin disease, Adams RM	1999	✓	
	45	Protective gloves for occupational use, Boman A	2005	✓	
	117	Handbook of occupational dermatology, Kanerva L	2000	✓	
	157	Surface and dermal monitoring for toxic exposures, Ness SA	1994	✓	
	229	Dermatotoxicology, Zhai H	2004	✓	
Brochure, pamphlet	60	A safety and health practitioner's guide to skin protection, CPWR	2000	✓	✓
	61	Physicians' alert for occupational contact dermatitis among construction workers, CPWR	2001	✓	
Guideline	47	Guideline for hand hygiene in health-care settings, CDC	2002	✓	
	113	American national standard for hand protection selection criteria, International Safety Equipment Association (ISEA)	2005	✓	
Journal article—primary	31	Effect of personal hygiene on blood lead levels of workers at a lead processing facility, Askin DP	1997	✓	

*See footnotes at end of table.

(Continued)

PROFESSIONAL AUDIENCE
Table 4F (Continued). Risk management

Resource type	ID	Title, author	Yr	Subtopics F.1*	F.2
Journal article—primary (Continued)	98	Survey assessment of worker dermal exposure and underlying behavioral determinants, Geer LA	2007	✓	
	123	Skin cleansers for occupational use: testing the skin compatibility of different formulations, Klotz A	2003	✓	
	131	Skin exposure to aliphatic polyisocyanates in the auto body repair and refinishing industry: a qualitative assessment, Liu Y	2007	✓	
	168	Risk assessment and exposure control in an occupational setting, Packham CL	1996	✓	✓
Journal article—review, meta-analysis	20	Skin protection programmes, Agner T	2002	✓	✓
	26	Cleansing without compromise: The impact of cleansers on the skin barrier and the technology of mild cleansing, Ananthapadmanabhan KP	2004	✓	
	30	Occupational contact dermatitis, Antezana M	2003	✓	
	38	Skin exposure to isocyanates: Reasons for concern, Bello D	2007	✓	
	42	In-use testing and interpretation of chemical-resistant glove performance, Boeniger MF	2002	✓	
	51	Concepts of skin protection: Considerations for the evaluation and terminology of the performance of skin protective equipment, Brouwer DH	2005	✓	
	53	Strategies for prevention: Occupational contact dermatitis, Brown T	2004	✓	✓
	69	Occupational issues of irritant contact dermatitis, Chew AL	2003	✓	
	73	Pesticide-related illness among migrant farm workers in the United States, Das R	2001	✓	

*See footnotes at end of table.

(Continued)

PROFESSIONAL AUDIENCE
Table 4F (Continued). Risk management

				Subtopics	
Resource type	ID	Title, author	Yr	F.1*	F.2
Journal article—review, meta-analysis (Continued)	78	The epidemiology of occupational contact dermatitis, Diepgen TL	1999	✓	
	79	Skin-conditioning products in occupational dermatology, Elsner P	2003	✓	
	103	Occupation-related allergies in dentistry, Hamann CP	2005	✓	
	106	Misinterpretation and misuse of exposure limits, Hewett P	2001		✓
	116	Dermatological aspects of a successful introduction and continuation of alcohol-based hand rubs for hygienic hand disinfection, Kampf G	2003	✓	
	122	A critique of assumptions about selecting chemical-resistant gloves: A case for workplace evaluation of glove efficacy, Klingner TD	2002	✓	
	125	Occupational skin-protection products—a review, Kresken J	2003	✓	
	127	Effectiveness of skin protection creams as a preventive measure in occupational dermatitis: A critical update according to criteria of evidence-based medicine, Kutting B	2003	✓	
	128	Toxicity of methyl methacrylate in dentistry, Leggat PA	2003	✓	
	135	Occupational contact dermatitis, Lushniak BD	2004	✓	
	140	Dermal toxicity due to industrial chemicals, Mathur AK	2002	✓	
	146	Differences between the sexes with regard to work-related skin disease, Meding B	2000	✓	
	163	A toolkit for dermal risk assessment and management: an overview, Oppl R	2003		✓

*See footnotes at end of table.

(Continued)

PROFESSIONAL AUDIENCE
Table 4F (Continued). Risk management

Resource type	ID	Title, author	Yr	Subtopics F.1*	F.2
Journal article—review, meta-analysis (Continued)	203	Management of dermatitis in the rubber manufacturing industry, Toeppen-Sprigg B	1999	✓	✓
	224	The dermal toxicity of cement, Winder C	2002	✓	
Other—Guideline from private lab	70	A guide to dermal exposure reduction, Colormetric Laboratories Inc.	1999	✓	✓
Technical publication/report	129	Epidemiology of skin and respiratory diseases among hairdressers, Leino T	2001	✓	
	152	NIOSH pocket guide to chemical hazards, NIOSH	2004	✓	
Web page	155	HSDB [Web page], NLM	2005	✓	
	208	Emergency response guidebook [Web page], United States Department of Transportation (USDOT)	2004	✓	
	220	Dermatitis: safety and health assessment and research for prevention (SHARP) [Home page] WADLI	2005	✓	
Web site	16	CPI [Web site], ACC	2006	✓	
	18	ATSDR [Home page], ATSDR	2005	✓	
	22	AAFP [Home page], AAFP	2005	✓	
	32	ASTM International [Home page], ASTM International	2006	✓	
	58	CCOHS [Home page], CCOHS	2005	✓	
	71	The pioneer in the reduction of dermal exposure [Colormetric Laboratories, Inc. Home page], Colormetric Laboratories, Inc.	2005	✓	
	82	Dermatological engineering [Web site], Enviroderm Services	2005	✓	✓

*See footnotes at end of table.

(Continued)

PROFESSIONAL AUDIENCE
Table 4F (Continued). Risk management

				Subtopics	
Resource type	ID	Title, author	Yr	F.1*	F.2
Web site (Continued)	87	USEPA [Home page], USEPA	2005	✓	
	105	Skin at work [Web site], HSE	2005	✓	
	112	ILO [Home page], ILO	2005	✓	
	153	NIOSH [Home page], NIOSH	2005	✓	
	154	TOXNET—Databases on toxicology, hazardous chemicals, environmental health, and toxic releases [Home page], NLM	2005	✓	
	162	OSHA [Home page], OSHA	2005	✓	
	164	OWIIPP [Web site], ORDHS	2005	✓	

*F.1=Exposure Control Strategies; F.2=Risk Assessment Protocols

CHAPTER 5
Overall Information Availability

5.1 Evaluation of Information Gaps

The *Indexed Dermal Bibliography* is a collection of information resources for the anticipation, recognition, evaluation, and control of occupational skin exposures to chemicals. It is not intended to provide an exhaustive list of all publications. Rather, it includes books, review papers, regulations, and databases available to the public and credible information available on the internet. The search for was limited to resources produced from 1995 and beyond and to those dealing specifically with occupational exposures to chemicals. None of the information sources was evaluated for the accuracy of the information presented.

Within the limits of this effort, some trends were seen in the availability of resources for each topic and subtopic. Additional primary research articles produced in 1995 or later and review articles or resources produced before 1995 may be available that were not considered in this evaluation.

Overall, several topic areas appeared to have limited information. These include:

- Information on conducting risk assessments for dermal exposure.
- Guidance on the interpretation of quantitative exposure assessments and on DOELs.
- Useable information on the effectiveness of dermal exposure control measures.
- Biological monitoring performance to assess the contribution of dermal exposure to overall exposures.
- Reviews of chemical mixtures and how different combinations of chemicals can affect exposures and health effects, as well as how to assess and manage exposures to chemical mixtures.
- Protocols for evaluating biological responses to mixtures.
- Reviews and guidance documents on assessing, both qualitatively and quantitatively, intervention effectiveness and intervention design.
- Reviews and guidance documents on how to implement control measures and on how to evaluate the effectiveness of control measures for dermal exposures.
- Brochures or other educational materials that address specific chemicals that are known skin hazards.

- Risk management strategies for particular chemicals or occupations and tasks. Construction and working with cement is an exception to this (see the eLCOSH Web site) and serves as occupations and particular chemical exposure types.
- Protocols and checklists for risk management, risk characterization, and surveillance.
- Exposure control plans designed specifically to prevent dermal hazards.

Below is a list of topics and subtopics for each audience for which few resources were identified.

General Audience

- "Specific chemicals." Information on specific chemicals was limited for the general audience, and this type of information is usually found in the professional audience references.
- There were few resources with quantitative descriptions on how exposure intensity and frequency influence exposure conditions.
- There were few guidance documents written for non-experts specific to skin hazards for qualitatively assessing skin exposure (e.g., protocols or checklists to characterize exposure when exposure data are not available, checklists and other tools for identifying exposures to hazardous chemicals, and checklists and other tools for identifying workplace conditions that contribute to skin exposure).
- There were few protocols and checklists specific to skin hazard exposure to use in identifying skin hazards at the workplace.
- There were few protocols and checklists for qualitatively identifying risks from skin exposure.
- There were few protocols and checklists for qualitatively monitoring potential skin exposures.

Professional Audience

- There are few checklists and questionnaires for quantifying skin exposure incidences.
- There is little available on how to assess the contribution of dermal exposures to overall exposures.
- More examples are needed of risk assessment for chemical mixtures.

- Resources are needed on how to develop a dermal exposure reduction goal based on the findings, both qualitative and quantitative, of a risk assessment.
- More guidance is needed on how to develop risk management and control measures in order to achieve dermal exposure reduction goals that consider the industrial process or tasks, regulatory requirements or guidance, or experiences in similar exposure situations.
- More resources are needed for the design, selection, and implementation of evaluation methods in order to demonstrate the effectiveness of intervention approaches for exposure control.

5.2 Future Research

Future research should consider the gaps presented above, and include (but not limited to):

General Audience

- Materials describing the concepts of exposure intensity and frequency with respect to skin hazards.
- Protocols or checklists for identifying skin hazards and qualitatively identifying risks and monitoring for skin exposures.

Professional Audience

- Development of educational materials on skin exposures and hazards for use in instruction and hazard communication.
- Development of guidance documents and examples of risk assessment of dermal hazards.
- Development of risk management approaches and controls for dermal exposure.
- Development of methods to evaluate the effectiveness of dermal hazard intervention approaches.

APPENDIX A
Full Resource Citations and Summaries

This appendix contains the full citations and summary information for all the resources referenced in the Indexed Dermal Bibliography, listed numerically by the assigned ID number.

Appendix A: Full Resource Citations and Summaries

Article ID:	1	
Citation:	Agency for Toxic Substance and Disease Registry (ATSDR) [1996]. Skin lesions and environmental exposures: an overview for the occupational health nurse. ATSDR. AAOHN J *44*(11): 529–540.	
Resource type:	Journal article—review, meta-analysis	
Educational materials:	No	
Number of references:	0	
Industries/occupations:		
Specific process:		
Chemical:		
Specific chemicals:		
Mixtures:	No	
Audience:	Professional	
Topics addressed:	A	Overview
	A.2	Health hazards resulting from skin exposure to chemicals
	A.4	Skin physiology and function as barriers to chemical insults
	B	Surveillance and clinical aspects
	B.4	Clinical protocols for recognition of skin exposure health effects
	C	Exposure Characterization
	C.4	Direct methods to measure exposure
	C.4.B	Skin
	D	Hazard identification
	D.1	Potential health effects resulting from specific chemicals
	D.1.A	Irritant contact dermatitis
	D.1.B	Allergic contact dermatitis/sensitization
	D.1.C	Systemic toxicity
	D.1.D	Other health effects
Summary:	This paper presents a detailed discussion of pathophysiology, etiologies, diagnosis, and treatment for seven skin conditions associated with environmental exposures: irritant contact dermatitis, allergic contact dermatitis, photosensitivity contact dermatitis, chloracne, pigment alterations, contact urticaria, and malignant neoplasms. There is also a discussion of a few diagnostic procedures including patch testing, photopatch testing, and skin biopsy.	

Appendix A: Full Resource Citations and Summaries

Article ID:	2
Citation:	United States Environmental Protection Agency (USEPA), Office of Prevention, Pesticides and Toxic Substances (OPPTS) [1998]. Health effects test guidelines: OPPTS 870.2500 acute dermal irritation. Washington, DC: USEPA, OPPTS.
Resource type:	Guideline
Educational materials:	No
Number of references:	14
Industries/occupations:	
Specific process:	
Chemical:	
Specific chemicals:	
Mixtures:	No
Audience:	Professional
Topics addressed:	D Hazard identification
	D.3 Characterization protocols
	D.3.A Corrosivity
	D.3.B Irritation potential
Summary:	This guideline specifies a procedure for testing acute dermal irritation of pesticides on animals. It is intended to meet testing requirements of both the Federal Insecticide, Fungicide, and Rodenticide Act (FIFRA) (7 U.S.C. 136, et seq.) and the Toxic Substances Control Act (TSCA) (15 U.S.C. 2601). The source materials used in developing this harmonized OPPTS test guideline are 40 CFR 798.4470 Primary Dermal Irritation; OPP 81-5 Primary Dermal Irritation (Pesticide Assessment Guidelines, Subdivision F-Hazard Evaluation; Human and Domestic Animals) USEPA report 540/09-82-025, 1982; and OECD 404 Acute Dermal Irritation/ Corrosion.

Article ID:	3
Citation:	National Institute for Occupational Safety and Health (NIOSH) [1998]. What you need to know about occupational exposure to metalworking fluids. Cincinnati, OH: U.S. Department of Health and Human Services (DHHS), Public Health Service (PHS), Centers for Disease Control and Prevention (CDC), NIOSH, DHHS (NIOSH) Publication No. 98–116.
Resource type:	Guideline
Educational materials:	Yes

Appendix A: Full Resource Citations and Summaries

Number of references:	96
Industries/occupations:	
Specific process:	
Chemical:	Heavy metals/inorganic compounds, petroleum products & lubricants
Specific chemicals:	
Mixtures:	No
Audience:	General

Topics addressed:	A	Overview
	A.2	Health hazards resulting from skin exposure to chemicals
	B	Exposure characterization
	B.2	Factors that influence exposure conditions
	B.2.B	Exposure controls
	E	Risk management
	E.3	"Best practices"/guidelines/recommendations
	E.3.B	Engineering controls
	E.3.C	Work practice/administration controls
	E.3.D	PPE and PPE rules
	E.3.E	Skin management, barrier creams, moisturizers, cleansers, and rubs

Summary:	This document summarizes the findings of the NIOSH Criteria Document: *NIOSH Criteria for a Recommended Standard: Occupational Exposure to Metalworking Fluids* (NIOSH Publication Number 98–102). It also provides a critical review of the scientific and technical information available on the subject as well as a scientific basis for the recommendations. It is an educational document intended to communicate basic information.

Article ID:	**4**
Citation:	Harris R, ed. [2000]. Occupational skin exposure—absorption of chemical agents and assessment of exposures. In: Patty's Industrial Hygiene, 5th ed. Vol. I, Recognition and evaluation of chemical agents. Indianapolis, IN: John Wiley & Sons.
Resource type:	Book/monograph, chapter
Educational materials:	No
Number of references:	142
Industries/occupations:	

Appendix A: Full Resource Citations and Summaries

Specific process:		
Chemical:		
Specific chemicals:		
Mixtures:	No	
Audience:	Professional	
Topics addressed:	A	Overview
	A.4	Skin physiology and function as barriers to chemical insults
	C	Exposure characterization
	C.2	Description of factors influencing exposure conditions
	C.2.E	Uptake
	D	Hazard identification
	D.1	Potential health effects resulting from specific chemicals
	D.1.A	Irritant contact dermatitis
	D.1.B	Allergic contact dermatitis/sensitization
	F	Risk management
	F.1	Exposure control strategies
	F.1.A	Substitution
	F.1.B	Engineering controls
	F.1.C	Work practice/Administrative controls
	F.1.D	PPE and PPE rules
	F.1.E	Skin management, barrier creams, moisturizers, cleansers, and rubs
Summary:	This chapter from *Patty's Industrial Hygiene Volume 1* includes a discussion of factors that cause and contribute to occupational dermatoses, covering chemicals, primary irritants, allergic contact dermatitis, plants and wood, photosensitivity, mechanical, physical, and biological factors. The chapter also discusses the physiology of the skin, patch tests, prevention, and control.	

Article ID:	5
Citation:	Keil CB, ed. [2000]. Dermal exposure modeling. In: Mathematical models for estimating occupational exposure to chemicals. Fairfax, VA: American Industrial Hygiene Association (AIHA).
Resource type:	Book/monograph, chapter
Educational materials:	No

Indexed Dermal Bibliography

Appendix A: Full Resource Citations and Summaries

Number of references:	30
Industries/occupations:	
Specific process:	
Chemical:	
Specific chemicals:	
Mixtures:	No
Audience:	Professional
Topics addressed:	C Exposure characterization
	C.2 Description of factors influencing exposure conditions
	C.2.B Exposure concentration
	C.2.C Skin area affected
	C.2.E Uptake
	C.5 Exposure modeling
Summary:	This chapter of the book focuses on estimating dermal exposures. There is a discussion of absorption mechanics, absorption factors, modeling, and data gaps and suggestions for additional research.

Article ID:	**6**
Citation:	Health and Safety Executive (HSE) [2000]. Selecting protective gloves for work with chemicals. Sudbury, UK: HSE Books.
Resource type:	Brochure, pamphlet
Educational materials:	Yes
Number of references:	3
Industries/occupations:	
Specific process:	
Chemical:	General—overview
Specific chemicals:	
Mixtures:	No
Audience:	General
Topics addressed:	E Risk management
	E.3 "Best practices"/guidelines/recommendations
	E.3.D PPE and PPE rules
Summary:	This leaflet, for employers and the self-employed, provides basic advice on selecting gloves to protect the wearer from chemical agents. It discusses UK law, chemical resistance of protective gloves, and selection of gloves. A PDF version of this leaflet is available on their Web site [www.hse.gov.uk/].

Appendix A: Full Resource Citations and Summaries

Article ID:	7
Citation:	Occupational Health Department [2000]. Did you know about the health hazards of benzene? Singapore: Republic of Singapore, Ministry of Manpower, Occupational Health Department.
Resource type:	Brochure, pamphlet
Educational materials:	Yes
Number of references:	0
Industries/occupations:	
Specific process:	
Chemical:	Solvents
Specific chemicals:	Benzene
Mixtures:	No
Audience:	General

Topics addressed:

A	Overview
A.1	Occurrence of skin exposures in the workplace
A.2	Health hazards resulting from skin exposure to chemicals
C	Hazard identification
C.1	Risk phrases, hazard symbols, skin designations
E	Risk management
E.3	"Best practices"/guidelines/recommendations
E.3.A	Substitution
E.3.B	Engineering controls
E.3.C	Work practice/administration controls
E.3.D	PPE and PPE rules

Summary: This brochure from the Singapore Department of Industrial Health presents the hazards of benzene. It describes properties, main uses, exposure hazards, acute effects (narcotic effect, drying effect on skin and mucous membranes), chronic effects (anemia, leukemia), technical control measures (substitution, engineering controls, personal protection), and medical control measures (preemployment examinations, periodic medical examinations).

Article ID:	8
Citation:	Occupational Health Department [2000]. Did you know the hazards of solvents? Singapore: Republic of Singapore, Ministry of Manpower, Occupational Health Department.

Indexed Dermal Bibliography

Appendix A: Full Resource Citations and Summaries

Resource type:	Brochure, pamphlet
Educational materials:	Yes
Number of references:	0
Industries/occupations:	
Specific process:	
Chemical:	Solvents
Specific chemicals:	
Mixtures:	No
Audience:	General

Topics addressed:		
	A	Overview
	A.1	Occurrence of skin exposures in the workplace
	A.2	Health hazards resulting from skin exposure to chemicals
	E	Risk management
	E.3	"Best practices"/guidelines/recommendations
	E.3.B	Engineering controls
	E.3.C	Work practice/administration controls

Summary: This leaflet from the Singapore Department of Industrial Health presents the hazards of solvents. It describes where solvents are used, why solvents are hazardous (volatility, flammability, explosivity, reactivity), their acute health effects (irritation of eyes, nose and throat, headache, nausea, poor coordination, arrhythmia), and their chronic health effects (skin dryness, allergic reactions, neurobehavioural changes, liver damage, paralysis, leukaemia). Other topics include storage and handling, limitation of exposure, safe working methods, medical supervision, and first-aid measures.

Article ID:	**9**
Citation:	HSE [2001]. Assessing and managing risks at work from skin exposure to chemical agents: guidance for employers and health and safety specialists. Sudbury, UK: HSE Books.
Resource type:	Brochure, pamphlet
Educational materials:	Yes
Number of references:	10
Industries/occupations:	General—overview
Specific process:	Lists occupational groups of concern including hairdressers.
Chemical:	General—overview

Appendix A: Full Resource Citations and Summaries

Specific chemicals:	methylene bis (discussed briefly)	
Mixtures:	No	
Audience:	General	
Topics addressed:	A	Overview
	A.1	Occurrence of skin exposures in the workplace
	A.2	Health hazards resulting from skin exposure to chemicals
	A.3	Dermal regulations and skin notations
	B	Exposure characterization
	B.1	Job/tasks, industries/processes, or chemicals associated with skin exposures
	C	Hazard identification
	C.3	Protocols/checklists to identify skin hazards in the workplace
	E	Risk management
	E.3	"Best practices"/guidelines/recommendations
	E.3.A	Substitution
	E.3.B	Engineering controls
	E.3.C	Work practice/administration controls
	E.3.D	PPE and PPE rules
Summary:	This guidance from the UK provides practical advice for employers and the self-employed to reduce the risk to skin exposures from chemicals. The guidance explains how health effects can be caused by skin exposure to chemicals, provides examples of chemicals known to cause health effects, and offers advice for assessing and reducing the risk of skin exposures.	

Article ID:	**10**
Citation:	USEPA [2001]. Risk assessment guidance for superfund (RAGS), Vol. I: Human health evaluation manual (Part E, supplemental guidance for dermal risk assessment). Washington, DC: USEPA.
Resource type:	Technical publication/report
Educational materials:	No
Number of references:	93
Industries/occupations:	Hazardous waste management
Specific process:	
Chemical:	

Indexed Dermal Bibliography

Appendix A: Full Resource Citations and Summaries

Specific chemicals:	Contains permeability coefficients for 28 inorganic compounds including 12 chromium compounds
Mixtures:	No
Audience:	Professional
Topics addressed:	C Exposure characterization
	C.2 Description of factors influencing exposure conditions
	C.2.A Exposure intensity/frequency/duration
	C.2.B Exposure concentration
	C.2.C Skin area affected
	C.2.E Uptake
	C.5 Exposure modeling
Summary:	The supplemental guidance section (Part E) to the *Risk Assessment Guidance for Superfund(RAGS), Volume I: Human Health Evaluation Manual* incorporates and updates the principles of the USEPA interim report, *Dermal Exposure Assessment: Principles and Applications*, released in 1992. Part E contains methods for conducting dermal risk assessments. Chapters include introduction and flowchart, hazard identification, exposure assessment, toxicity assessment, risk characterization, and conclusion/recommendations.

Article ID:	**11**
Citation:	HSE [2001]. Cost and effectiveness of chemical protective gloves for the workplace. Sudbury, UK: HSE Books.
Resource type:	Brochure, pamphlet
Educational materials:	Yes
Number of references:	8
Industries/occupations:	
Specific process:	
Chemical:	
Specific chemicals:	
Mixtures:	No
Audience:	General
Topics addressed:	E Risk management
	E.3 "Best practices"/guidelines/recommendations
	E.3.D PPE and PPE rules
Summary:	This UK publication provides employers with advice on the cost and effectiveness of gloves and personal protective

Appendix A: Full Resource Citations and Summaries

equipment (PPE) for those industries where employees incur risk from dermal exposure to chemicals. It is available for purchase on the HSE Web site [www.hse.gov.uk/].

Article ID:	**12**
Citation:	HSE [2001]. Choice of skin care products for the workplace. Sudbury, UK: HSE Books.
Resource type:	Brochure, pamphlet
Educational materials:	Yes
Number of references:	0
Industries/occupations:	
Specific process:	
Chemical:	
Specific chemicals:	
Mixtures:	No
Audience:	General
Topics addressed:	A Overview
	A.2 Health hazards resulting from skin exposure to chemicals
	E Risk management
	E.3 "Best practices"/guidelines/recommendations
	E.3.E Skin management, barrier creams, moisturizers, cleansers, and rubs
Summary:	This UK publication provides employers with information on skin care products, their selection, and how they fit into an overall skin care program for those industries where employees incur risk from dermal exposure to chemicals. This is available for purchase on the HSE Web site [www.hse.gov.uk/].

Article ID:	**13**
Citation:	American Skin Association (ASA) [2005]. [www.americanskin.org/frameset.htm].
Resource type:	Web site
Educational materials:	Yes
Number of references:	
Industries/occupations:	General—overview
Specific process:	

Appendix A: Full Resource Citations and Summaries

Chemical:	General—overview, latex	
Specific chemicals:		
Mixtures:	No	
Audience:	General	
Topics addressed:	A	Overview
	A.1	Occurrence of skin exposures in the workplace
	A.2	Health hazards resulting from skin exposure to chemicals
Summary:	The ASA is a volunteer-led health organization that works on issues associated with skin disorders. One of the primary missions of the ASA is to raise public awareness of the wide range of skin disorders through ongoing public education. They produce a consumer-oriented quarterly newsletter called *Skin Facts*. Past issues can be accessed through the Web site's archives. Past articles have included skin disease in the workplace and latex sensitivity.	

Article ID:	**14**	
Citation:	Montana Department of Labor and Industries (MTDLI)—Employment Relations [2005]. [http://erd.dli.mt.gov/].	
Resource type:	Web site	
Educational materials:	No	
Number of references:		
Industries/occupations:	General—overview	
Specific process:		
Chemical:	General—overview, latex	
Specific chemicals:		
Mixtures:	No	
Audience:	General	
Topics addressed:	A	Overview
	A.1	Occurrence of skin exposures in the workplace
	A.2	Health hazards resulting from skin exposure to chemicals
	B	Exposure characterization
	B.1	Job/tasks, industries/processes, or chemicals associated with skin exposures
	E	Risk management
	E.1	Overview of skin exposure control options
	E.3	"Best practices"/guidelines/recommendations

Appendix A: Full Resource Citations and Summaries

	E.3.A	Substitution
	E.3.B	Engineering controls
	E.3.C	Work practice/administration controls
	E.3.D	PPE and PPE rules
	E.3.E	Skin management, barrier creams, moisturizers, cleansers, and rubs
	E.4	Guidelines/recommendations for postexposure skin decontamination
Summary:		The MT DLI Safety and Health Bureau is responsible for workplace safety and health through inspection, consultation, technical assistance, and training. Over 100 different occupational safety and health documents and brochures are available either electronically through their Web site or hard copies can be ordered through their Web site. Documents are accessed through the program samples topic on the Health and Safety Bureau's drop-down box. Dermal exposure-related documents and brochures found at this site include:

- *Dermatitis Prevention: Occupational Skin Disorders*
- *Latex Allergy*
- *Job Safety Analysis Packet* (though generic, can be used to evaluate dermal hazards)

Article ID:	15	
Citation:	1105 Media, Inc. [2006]. Occupational health and safety. [www.ohsonline.com].	
Resource type:	Web site	
Educational materials:	No	
Number of references:		
Industries/occupations:	General—overview, Medical Services	
Specific process:		
Chemical:	General—overview, heavy metals/inorganic compounds, latex, plastics, resins	
Specific chemicals:	Hexavalent chromium (CrVI)	
Mixtures:	No	
Audience:	General	
Topics addressed:	A	Overview
	A.1	Occurrence of skin exposures in the workplace
	A.3	Dermal regulations and skin notations
	B	Exposure characterization

Appendix A: Full Resource Citations and Summaries

	B.1	Job/tasks, industries/processes, or chemicals associated with skin exposures
	B.3	Protocols/checklists to characterize exposure to skin hazards
	C	Hazard identification
	C.1	Risk phrases, hazard symbols, skin designations
	C.3	Protocols/checklists to identify skin hazards in the workplace
	E	Risk management
	E.2	Protocols/checklists to monitor potential exposures
	E.3	"Best practices"/guidelines/recommendations
	E.3.C	Work practice/administration controls
	E.3.D	PPE and PPE rules
	E.3.E	Skin management, barrier creams, moisturizers, cleansers, and rubs
Summary:		The *Occupational Health and Safety* online magazine periodically features articles on dermal exposure and control. They have archived past issues that can be searched by subject. Dermal exposure articles that can be found online include:

- "Effective Dermal Protection"
- "Butyl & Viton Hand Protection"
- "Latex Allergy & Dermatitis"
- "CrVI: New Regulations and Detection Methods"

Article ID:	16	
Citation:	American Chemistry Council (ACC) [2006]. Center for the Polyurethanes Industry (CPI). [www.polyurethane.org/].	
Resource type:	Web site	
Educational materials:	Yes	
Number of references:		
Industries/occupations:	Manufacturing—Chemical	
Specific process:		
Chemical:	Plastics and resins	
Specific chemicals:	Diisocyanates	
Mixtures:	No	
Audience:	General	
Topics addressed:	A	Overview

Appendix A: Full Resource Citations and Summaries

A.1	Occurrence of skin exposures in the workplace
A.2	Health hazards resulting from skin exposure to chemicals
B	Exposure characterization
B.1	Job/tasks, industries/processes, or chemicals associated with skin exposures
B.2	Factors that influence exposure conditions
B.2.A	Exposure intensity/frequency
B.2.B	Exposure controls
E	Risk management
E.1	Overview of skin exposure control options
E.3	"Best practices"/guidelines/recommendations
E.3.A	Substitution
E.3.B	Engineering controls
E.3.C	Work practice/administration controls
E.3.D	PPE and PPE rules
E.4	Guidelines/recommendations for postexposure skin decontamination

Summary: The CPI is a business unit of the American Plastics Council. CPI's 84 members include U.S. producers or distributors of chemicals and equipment used to make polyurethane and polyurethane product manufacturers. Their Web site contains information on polyurethane health and safety. Resources of interest associated with dermal exposure issues include:

- Hyper-reactivity and Other Health Effects of Diisocyanates: Guidelines for Medical Personnel
- Working with TDI: What you should know
- Polyol Resin Blends Safety and Handling Guidelines
- Working with MDI: What you should know

Article ID:	**17**
Citation:	Adams RM [1999]. Occupational skin disease, 3rd ed. Philadelphia: Saunders.
Resource type:	Book/monograph, whole
Educational materials:	No
Number of references:	5112
Industries/occupations:	General—overview, Agricultural, Beauty/Cosmetology, Cleaning/Janitorial/Maid, Construction, Forestry/Fisheries, Manufacturing—Chemical, Manufacturing—Other,

Indexed Dermal Bibliography

Appendix A: Full Resource Citations and Summaries

	Medical Services, Service—Food, Service—Medical, Service—Other, Transportation/Communications/Utility
Specific process:	Describes the occupation and risks, lists irritants, standard allergens and additional allergens for the following occupations:
	Air hammer operators, abattoir workers, aircraft workers, anodizers, artists, asphalt workers, athletes, automobile mechanics, bakers, barbers, bartenders, bath attendants, battery makers, beekeepers, blueprint makers, bookbinders, brake lining workers, butchers, poultry processors, cabinet makers, candle makers, cannery workers, carpenters, cashiers, caulkers, cement workers, ceramic workers, chemists, cigarette and cigar makers, construction workers, cosmetologists, dairy workers, dentists and dental personnel, dry cleaners, electricians, electron microscopy workers, electroplaters, embalmers, engravers, firefighters, floor layers, florists, food service workers, forest workers and loggers and foresters, foundry workers, fur processors, glaziers, healthcare workers, highway construction workers, histology technicians, house workers, insulation workers, jewelers, laundry workers, locksmiths, machinists, medical personnel, metal polishers, musicians, office workers, optical technicians, painters and paperhangers, papermakers, performing artists, pest control workers, pharmacists, photographers, plastics assemblers and fabricators, plumbers and pipe fitters, police officers and detectives, postal workers, pottery and porcelain makers, printers, radio and television repairers, railroad shop workers, roofers, semiconductor and electronics workers, sheet-metal workers, shoe repairers, silk-screening workers, solderers and brazers, stonemasons, swimming pool personnel, tannery workers, tattoo artists, taxidermists, textile workers, theatrical artists, tile setters, tobacco workers, veterinarians, welders, wine makers, and wire drawing operators.
Chemical:	General—overview, heavy metals/inorganic compounds, pesticides, petroleum products & lubricants, plastics and resins, rubber additives, soaps and detergents, solvents, other: semiconductors, plants, steroids, paints
Specific chemicals:	
Mixtures:	Yes
Audience:	Professional
Topics addressed:	C Exposure characterization
	C.2 Description of factors influencing exposure conditions
	C.2.A Exposure intensity/frequency/duration
	C.2.B Exposure concentration

Appendix A: Full Resource Citations and Summaries

C.2.C	Skin area affected
C.2.E	Uptake
C.4	Direct methods to measure exposure
C.4.A	Surfaces
C.4.B	Skin
C.4.C	Biomonitoring
C.5	Exposure modeling
D	Hazard identification
D.1	Potential health effects resulting from specific chemicals
D.1.A	Irritant contact dermatitis
D.1.B	Allergic contact dermatitis/sensitization
D.1.C	Systemic toxicity
D.1.D	Other health effects
D.3	Characterization protocols
D.3.E	Measurement of skin permeation rates and reservoir effects
F	Risk management
F.1	Exposure control strategies
F.1.E	Skin management, barrier creams, moisturizers, cleansers, and rubs

Summary: This comprehensive book is a standard reference on occupational skin diseases. It provides an overview as well as an in-depth discussion of skin diseases associated with dozens of specific occupations, their causes, and health effects. Each chapter was written by national and international researchers and is individually referenced. It includes a step-by-step guide for making precise diagnoses, considerations for differential diagnoses, and practical solutions for skin disease problems.

CHAPTER HEADINGS: Irritants/Allergic Contact Dermatitis:

General Principles and Causes/Physical Causes of Occupational Skin Disease/Systemic Toxicity from Percutaneous Absorption/Biological Causes/Contact Urticaria Due to Occupational Exposure/Acne and Folliculitis Caused By Mechanical Factors "Chloracne"/Occupational Skin Cancer/Occupational Nail Disorders/Phototoxicity and Photosensitivity Reactions /Occupational Nail Disorders/Diagnosis and Differential Diagnosis/Atopy, Atopic Dermatitis and Occupational Skin Disease/Diagnostic Patch Testing/The Computer in Occupational Skin Disease/Multiple Chemical Sensitivities/

Appendix A: Full Resource Citations and Summaries

Prevention, Rehabilitation, Treatment/Health Risk Assessment and Occupational Dermatology/Workers Compensation/Plant Inspection/Industrial Processes Commonly Associated with Skin Disease/Soaps and Detergents/Cosmetics/Corticosteroids/Metals/Plastics and Platicizers/Semiconductor Industry/Paints, Varnishes and Lacquers/Solvents/Occupational Skin Problems from Natural and Synthetic Rubber/Petroleum and Petroleum Derivatives/Occupational Dermatitis from Plants and Woods/Pesticides and Other Agriculture Chemicals/Job Descriptions with Their Irritants and Allergens.

Article ID:	**18**
Citation:	ATSDR [2008]. [www.atsdr.cdc.gov/].
Resource type:	Web site
Educational materials:	No
Number of references:	
Industries/occupations:	General—overview
Specific process:	
Chemical:	Abrasives, cleaning agents, coolants, corrosives, fiberglass and other fibers, heavy metals/inorganic compounds, nanoparticles, organic dyes, particulates, pesticides, petroleum products & lubricants, PAHs, PCBs, rubber additives, solvents
Specific chemicals:	Information on hundreds of chemicals
Mixtures:	No
Audience:	Professional
Topics addressed:	A Overview
	A.1 Occurrence of skin exposures in workplace
	B Surveillance and clinical aspects
	B.4 Clinical protocols for recognition of skin exposure health effects
	D Hazard identification
	D.1 Potential health effects resulting from specific chemicals
	D.1.A Irritant contact dermatitis
	D.1.B Allergic contact dermatitis/sensitization
	D.1.C Systemic toxicity
	D.1.D Other health effects
	D.1.E Contribution to overall exposure

Appendix A: Full Resource Citations and Summaries

	D.2	Summaries of health effects, dose-response relationships
	F	Risk management
	F.1	Exposure control strategies
	F.1.C	Work practice/Administrative controls
	F.1.D	PPE and PPE rules

Summary: The ATSDR is a federal public health agency of the U.S. DHHS. ATSDR is directed by a congressional mandate to perform specific functions concerning the effect on public health of hazardous substances in the environment. These functions include response to emergency releases of hazardous substances, information development and dissemination, and education and training concerning hazardous substances. The Web site contains a number of resources applicable to occupational dermal exposure to chemicals, including:

- Medical management guidelines (MMGs): Guidelines for acute chemical exposures were developed by ATSDR to aid emergency department physicians and other emergency healthcare professionals who manage acute exposures resulting from chemical incidents. Information provided in the guidelines includes potential routes of exposure, applicable exposure standards and guidelines, health effects, and protective measures to be taken by rescue workers, decontamination procedures and printable follow-up documents for patients who have been exposed to specified chemicals. Guidelines are available for approximately 50 chemicals. [http://www.atsdr.cdc.gov/MHMI/mmg.html]

- Toxicological profile information sheets: Toxicological profiles for hazardous substances found at National Priorities List (NPL) sites. Profiles are available on over 250 chemicals (see ID 19 for more details).

- ToxFAQs: A series of summaries taken from toxicological profiles and public health statements. Each fact sheet serves as a quick and easy-to-understand guide. Answers are provided to the most frequently asked questions (FAQs) about exposure to hazardous substances found around hazardous waste sites and the effects of exposure on human health.

- Interaction profiles for toxic substances: A series of documents being developed for certain priority mixtures that are of special concern to ATSDR. The purpose of the interaction profiles is to evaluate data on the toxicology of the mixture and on the joint toxic action of the chemicals in the mixture in

Appendix A: Full Resource Citations and Summaries

order to recommend approaches for exposure-based assessment of the potential hazard to public health.

Article ID:	**19**
Citation:	ATSDR [2005]. Toxicological profile information sheet. [www.atsdr.cdc.gov/toxpro2.html].
Resource type:	Web page
Educational materials:	No
Number of references:	
Industries/occupations:	General—overview
Specific process:	
Chemical:	Abrasives, cleaning agents, coolants, corrosives, hand cleansers, heavy metals/inorganic compounds, organic dyes, particulates, pesticides, petroleum products & lubricants, plastics and resins, PAHs, PCBs, rubber additives, solvents
Specific chemicals:	250 chemicals listed
Mixtures:	No
Audience:	Professional
Topics addressed:	A Overview
	A.2 Health hazards resulting from skin exposure to chemicals
	A.3 Investigation, intervention, and control of occupational skin exposures
	C Exposure characterization
	C.4 Direct methods to measure exposure
	C.4.C Biomonitoring
	D Hazard identification
	D.1 Potential health effects resulting from specific chemicals
	D.1.A Irritant contact dermatitis
	D.1.B Allergic contact dermatitis/sensitization
	D.1.C Systemic toxicity
	D.1.D Other health effects
	D.1.E Contribution to overall exposure
	D.2 Summaries of health effects, dose-response relationships
Summary:	The ATSDR produces toxicological profile information sheets for hazardous substances found at National

Appendix A: Full Resource Citations and Summaries

Priorities List (NPL) hazardous waste sites. Although geared toward environmental rather than occupational exposures, the sheets contain useful information for occupational settings as well. So far, 282 toxicological profiles have been published or are under development and cover more than 250 substances. Each chemical profile contains information on health effects, chemical and physical information, potential for human exposure, analytical methods, and regulations and advisories.

Each profile is written for a general audience and contains a public health statement that includes information written in nontechnical terms on what the chemical is, how one might be exposed to it, how the chemical enters and leaves the body, the effects of exposure, and medical tests to determine if a worker has been exposed.

Article ID:	20	
Citation:	Agner T, Held E [2002]. Skin protection programmes. Contact Dermatitis 47(5):253–56.	
Resource type:	Journal article—review, meta-analysis	
Educational materials:	No	
Number of references:	46	
Industries/occupations:		
Specific process:		
Chemical:		
Specific chemicals:		
Mixtures:	No	
Audience:	Professional	
Topics addressed:	F	Risk management
	F.1	Exposure control strategies
	F.1.C	Work practice/Administrative controls
	F.1.D	PPE and PPE rules
	F.1.E	Skin management, barrier creams, moisturizers, cleansers, and rubs
	F.2	Protocols for risk management
	F.2.B	Development of approach to achieve exposure reduction goal
Summary:	The article discusses 10 recommendations for skin protection involving washing, gloves, moisturizers, and behavioral changes.	

Appendix A: Full Resource Citations and Summaries

Article ID:	21	
Citation:	American Academy of Family Physicions (AAFP) [2004]. Skin problems: how to protect yourself from job-related skin problems. [http://familydoctor.org/750.xml].	
Resource type:	Web page	
Educational materials:	No	
Number of references:		
Industries/occupations:	General—overview	
Specific process:		
Chemical:	General—overview	
Specific chemicals:		
Mixtures:	No	
Audience:	General	
Topics addressed:	A	Overview
	A.1	Occurrence of skin exposures in the workplace
	A.2	Health hazards resulting from skin exposure to chemicals
	E	Risk management
	E.1	Overview of skin exposure control options
Summary:	This very general review of job-related skin problems includes a discussion on how workers can protect themselves against workplace skin hazards.	

Article ID:	22	
Citation:	AAFP [2005]. [www.aafp.org]	
Resource type:	Web site	
Educational materials:	No	
Number of references:		
Industries/occupations:	General—overview	
Specific process:		
Chemical:	General—overview	
Specific chemicals:		
Mixtures:	No	
Audience:	General	
Topics addressed:	A	Overview
	A.1	Occurrence of skin exposures in the workplace
	A.2	Health hazards resulting from skin exposure to chemicals

Appendix A: Full Resource Citations and Summaries

	B	Exposure characterization
	B.1	Job/tasks, industries/processes, or chemicals associated with skin exposures
	C	Hazard identification
	C.2	Tables/charts/lists of hazards for specific chemicals
	E	Risk management
	E.1	Overview of skin exposure control options
	E.3	"Best practices"/guidelines/recommendations
	E.3.D	PPE and PPE rules
	E.3.E	Skin management, barrier creams, moisturizers, cleansers, and rubs

Summary: The AAFP is one of the largest national medical organizations, representing more than 94,000 family physicians, family medicine residents, and medical students nationwide. The Web site contains information related to dermal exposure and occupational skin disease. Key information includes:

- Skin problems on the job-patient information handsheet—This handsheet, written for a more general audience, provides a brief overview of skin hazards and what workers can do to protect themselves.

- Occupational Skin Disease—This article, written for medical professionals, provides an overview of cause, diagnosis, and control of occupational skin disease, including a more detailed description of irritant contact dermatitis and allergic contact dermatitis and occupational groups at risk.

Article ID:	**23**
Citation:	American Conference of Governmental Industrial Hygienists (ACGIH) [2001]. Documentation of the threshold limit values for chemical substances, 7th ed. Cincinnati: ACGIH.
Resource type:	Guideline
Educational materials:	No
Number of references:	
Industries/occupations:	General—overview
Specific process:	
Chemical:	Abrasives, cleaning agents, coolants, corrosives, fiberglass and other fibers, heavy metals/inorganic compounds, latex, nanoparticles, organic dyes, particulates, pesticides,

Appendix A: Full Resource Citations and Summaries

	petroleum products & lubricants, plastics and resins, PAHs, PCBs, rubber additives, solvents
Specific chemicals:	Includes over 500 chemicals
Mixtures:	No
Audience:	Professional
Topics addressed:	D Hazard identification
	D.1 Potential health effects resulting from specific chemicals
	D.1.A Irritant contact dermatitis
	D.1.B Allergic contact dermatitis/sensitization
	D.1.C Systemic toxicity
	D.1.D Other health effects
	D.1.E Contribution to overall exposure
	D.2 Summaries of health effects, dose-response relationships
Summary:	ACGIH is a scientific association with a number of technical committees that develop professional practice guidelines, such as threshold limit values (TLVs) for chemical substances and physical agents and the biological exposure indices (BEIs) for selected chemicals. The TLVs and BEIs are developed as guidelines to assist in the control of health hazards. The documentation of the TLVs and BEIs provides the basic rationale for the development of TLVs and of BEIs. The publication consists of documentation for over 750 chemical and physical agents. These recommendations or guidelines are intended for use in the practice of industrial hygiene, to be interpreted and applied only by a person trained in this discipline. They are not developed for use as legal standards and ACGIH does not advocate their use as such. The documentation is available in hard copy, on CD, or downloadable from the ACGIH Web site.

Article ID:	24
Citation:	AIHA [2005]. 2005 Emergency response planning guidelines ERPG and workplace environmental exposure level WEELs handbook. Fairfax, VA: AIHA.
Resource type:	Brochure, pamphlet
Educational materials:	Yes
Number of references:	15
Industries/occupations:	
Specific process:	
Chemical:	

Appendix A: Full Resource Citations and Summaries

Specific chemicals:	
Mixtures:	No
Audience:	General
Topics addressed:	A Overview
	A.3 Dermal regulations and skin notations
Summary:	This pocket-sized emergency reference guide presents an overview of two sets of exposure limits: the AIHA ERPG (114) and WEELs (108). It contains recommended values for each series. In addition to a glossary, both ERPG and WEELs sections include background information, user guidance, value rationale, sample documents, and values. There is also an explanation on biological environmental exposure limits (BEELs).

Article ID:	**25**
Citation:	AIHA [2006]. [www.aiha.org].
Resource type:	Web site
Educational materials:	No
Number of references:	
Industries/occupations:	
Specific process:	
Chemical:	General—overview
Specific chemicals:	
Mixtures:	No
Audience:	Professional
Topics addressed:	A Overview
	A.5 Dermal regulations and skin notations
Summary:	The AIHA is a nonprofit, international association of occupational and environmental health professionals. Among other things, the AIHA Web site is a source of information on occupational and environmental health and safety topics and publications, including dermal exposure. AIHA members can serve on a number of committees that support AIHA's mission to promote healthy and safe environments by advancing the science, principles, practice, and value of industrial hygiene and occupational and environmental health and safety. The Dermal Project Team of the Exposure Assessment Strategies Committee focuses on issues associated with dermal exposure assessment. On the Dermal Project Team Web page are resources related to dermal exposure, including a list of general sources of dermal information.

Appendix A: Full Resource Citations and Summaries

The AIHA Workplace Environmental Exposure Levels (WEELs) Committee works on establishing and updating AIHA's WEELs. These include a skin designation for chemicals in which significant amounts may be absorbed through the skin, and therefore contribute to overall exposures.

Article ID:	**26**
Citation:	Ananthapadmanabhan KP, Moore DJ, Subramanyan K, Misra M, Meyer F [2004]. Cleansing without compromise: The impact of cleansers on the skin barrier and the technology of mild cleansing. Dermatol Ther *17*(1):16–25.
Resource type:	Journal article—review, meta-analysis
Educational materials:	No
Number of references:	40
Industries/occupations:	
Specific process:	
Chemical:	Hand cleansers
Specific chemicals:	
Mixtures:	No
Audience:	Professional
Topics addressed:	D Hazard identification
	D.1 Potential health effects resulting from specific chemicals
	D.1.A Irritant contact dermatitis
	F Risk management
	F.1 Exposure control strategies
	F.1.E Skin management, barrier creams, moisturizers, cleansers, and rubs
Summary:	This review discusses the benefits and health impacts of skin cleansers and compares different kinds of skin cleansing products.

Article ID:	**27**
Citation:	Andersen KE [2003]. Occupational issues of allergic contact dermatitis. Int Arch Occup Environ Health *76*(5):347–50.
Resource type:	Journal article—review, meta-analysis
Educational materials:	No
Number of references:	28

Appendix A: Full Resource Citations and Summaries

Industries/occupations:		
Specific process:		
Chemical:		
Specific chemicals:		
Mixtures:	No	
Audience:	Professional	
Topics addressed:	B	Surveillance and clinical aspects
	B.1	Surveillance study reporting incidences of occupational skin exposures
	B.1.A	Skin exposure major focus
	B.4	Clinical protocols for recognition of skin exposure health effects
	C	Exposure characterization
	C.4	Direct methods to measure exposure
	C.4.B	Skin
	C.4.C	Biomonitoring
	D	Hazard identification
	D.2	Summaries of health effects, dose-response relationships
Summary:	This review addresses occupational allergic contact dermatitis. The article discusses epidemiological data, diagnosis, exposure assessment, and dose-response relationship. Preventive measures are also discussed, though in general terms.	

Article ID:	**28**
Citation:	Andersen, KE [1999]. Systemic toxicity from percutaneous absorption. In: Adams RM, ed. Occupational skin disease. Philadelphia: Saunders, 69–85.
Resource type:	Book/monograph, chapter
Educational materials:	No
Number of references:	147
Industries/occupations:	
Specific process:	
Chemical:	Heavy metals/inorganic compounds, pesticides, rubber additives, solvents, phosphate esters, chlorinated hydrocarbons, topical drugs and toiletries, pharmaceuticals
Specific chemicals:	
Mixtures:	No

Indexed Dermal Bibliography

Appendix A: Full Resource Citations and Summaries

Audience:	Professional	
Topics addressed:	A	Overview
	A.4	Skin physiology and function as barriers to chemical insults
	B	Surveillance and clinical aspects
	B.1	Surveillance study reporting incidences of occupational skin exposures
	B.1.A	Skin exposure major focus
	B.1.B	Skin exposure minor focus
	C	Exposure characterization
	C.1	Workplace factors associated with harmful skin exposures
	C.2	Description of factors influencing exposure conditions
	C.2.A	Exposure intensity/frequency/duration
	C.2.B	Exposure concentration
	C.2.C	Skin area affected
	C.2.E	Uptake
	D	Hazard identification
	D.1	Potential health effects resulting from specific chemicals
	D.1.B	Allergic contact dermatitis/sensitization
	D.1.C	Systemic toxicity
Summary:	This comprehensive reference by over 40 clinician contributors discusses diagnosis, treatment, and prevention of occupational skin disease. This chapter addresses uptake, biotransformation, exposure/reaction patterns, and effects of specific chemicals.	

Article ID:	**29**
Citation:	Ansell Chemsafe [2005]. Ansell Chemsafe. [www.ansellchemsafe.com/Default.aspx].
Resource type:	Web site
Educational materials:	No
Number of references:	
Industries/occupations:	General—overview
Specific process:	
Chemical:	General—overview
Specific chemicals:	

106   *Indexed Dermal Bibliography*

Appendix A: Full Resource Citations and Summaries

Mixtures:	No
Audience:	General
Topics addressed:	A Overview
	A.1 Occurrence of skin exposures in the workplace
	E Risk management
	E.1 Overview of skin exposure control options
	E.3 "Best practices"/guidelines/recommendations
	E.3.D PPE and PPE rules
Summary:	Ansell is an Australian chemical glove manufacturer. Their Web site includes general information on dermal exposure to chemicals and protecting the skin as well as a software program, SpecwareTM, that can be used to assist in the glove selection process. Specware provides the user with glove use recommendations for a variety of commonly used chemicals. The information can be accessed through the software program on their Web site or with their hardcopy Specware guide which is available on request.

Article ID:	**30**
Citation:	Antezana M, Parker F [2003]. Occupational contact dermatitis. Immunol Allergy Clin North Am *23*(2):269–90.
Resource type:	Journal article—review, meta-analysis
Educational materials:	No
Number of references:	43
Industries/occupations:	Agricultural, Beauty/Cosmetology, Construction, Forestry/Fisheries, Manufacturing—Other, Service—Medical
Specific process:	Painting Printing Forestry Electronics
Chemical:	heavy metals/inorganic compounds, organic dyes, pesticides, rubber additives, other: adhesives, paints
Specific chemicals:	paraphenylenediamine, nickel, chromium, ethylenediamine, thimerosal
Mixtures:	No
Audience:	Professional
Topics addressed:	A Overview
	A.2 Health hazards resulting from skin exposure to chemicals
	B Surveillance and clinical aspects

Appendix A: Full Resource Citations and Summaries

B.4	Clinical protocols for recognition of skin exposure health effects
C	Exposure characterization
C.4	Direct methods to measure exposure
C.4.B	Skin
D	Hazard identification
D.1	Potential health effects resulting from specific chemicals
D.1.A	Irritant contact dermatitis
D.1.B	Allergic contact dermatitis/sensitization
F	Risk management
F.1	Exposure control strategies
F.1.D	PPE and PPE rules
F.1.E	Skin management, barrier creams, moisturizers, cleansers, and rubs

Summary: This paper presents the epidemiology, pathophysiology, and symptomology of occupational dermatitis, as well as diagnostic tests for dermatitis. There is some discussion of at-risk occupations, common allergens and irritants, and preventive management.

Article ID:	**31**
Citation:	Askin DP, Volkmann M [1997]. Effect of personal hygiene on blood lead levels of workers at a lead processing facility. AIHA J 58(10):752–53.
Resource type:	Journal article—primary
Educational materials:	No
Number of references:	3
Industries/occupations:	Waste management
Specific process:	
Chemical:	Heavy metals/inorganic compounds
Specific chemicals:	lead
Mixtures:	No
Audience:	Professional
Topics addressed:	

C	Exposure characterization
C.1	Workplace factors associated with harmful skin exposures
C.4	Direct methods to measure exposure
C.4.B	Skin

Appendix A: Full Resource Citations and Summaries

	F	Risk management
	F.1	Exposure control strategies
	F.1.C	Work practice/Administrative controls
Summary:	At a lead processing plant, lead was measured for 24 workers who were confident that their hands were clean. Samples were obtained by cleaning one hand with a wipe. Workers with more than one year's experience had a significantly positive correlation between lead on their hand and blood lead level, suggesting that lead on the skin enters the bloodstream. The route of entry was not investigated.	

Article ID:	32
Citation:	ASTM International [2006]. [www.astm.org/].
Resource type:	Web site
Educational materials:	No
Number of references:	
Industries/occupations:	General—overview
Specific process:	
Chemical:	Coolants, plastics and resins
Specific chemicals:	Isocyanates, metalworking fluids
Mixtures:	No
Audience:	Professional
Topics addressed:	

	C	Exposure characterization
	C.1	Workplace factors associated with harmful skin exposures
	C.4	Direct methods to measure exposure
	C.4.B	Skin
	D	Hazard identification
	D.1	Potential health effects resulting from specific chemicals
	D.1.B	Allergic contact dermatitis/sensitization
	F	Risk management
	F.1	Exposure control strategies
	F.1.A	Substitution
	F.1.B	Engineering controls
	F.1.C	Work practice/Administrative controls
	F.1.D	PPE and PPE rules

Indexed Dermal Bibliography

Appendix A: Full Resource Citations and Summaries

Summary: ASTM International is a voluntary standards development organization. Standards developed at ASTM are the work of ASTM members. These technical experts represent producers, users, consumers, government, and academia from over 100 countries. Standards and guides are available for sale on their Web site.

Some of the ASTM standards and guides relevant to dermal exposure include:

- E1497-05 *Standard Practice for Safe Use of Water-Miscible Metal Removal Fluids*—This practice provides guidelines for the selection and safe use of metal removal fluids, additives, and antimicrobials. This includes product selection, storage, dispensing, and maintenance.

- Standard E 1302, *Guide for Acute Animal Toxicity Testing of Water-Miscible Metalworking Fluids*—This guide defines acute animal toxicity tests and presents references for procedures that assess the acute toxicity of water-miscible metalworking fluid concentrates as manufactured.

- WK8210 *Standard Guide to Test Methods for Personal Protective Equipment Intended for Homeland Security Applications*—This guide provides a listing of test methods for personal protective equipment (PPE) intended to protect first and second responders, casualty receivers, and remediation personnel involved in homeland security incidents.

- STP 1408 *ISOCYANATES: Sampling, Analysis, and Health Effects*—11 peer-reviewed papers on topics such as, isocyanate determination in atmospheres; sampling strategy and control; and personal protective equipment.

- F1296-03 *Standard Guide for Evaluating Chemical Protective Clothing*—This guide is intended to promote the proper selection, use, maintenance, and understanding of the limitations of chemical protective clothing by users, employers, employees, and other persons involved in programs requiring CPC.

Article ID: 33

Appendix A: Full Resource Citations and Summaries

Citation:	Balsat A, De Graeve J, Mairiaux P [2003]. A structured strategy for assessing chemical risks, suitable for small and medium-sized enterprises. Ann Occup Hyg *47*(7): 549–56.
Resource type:	Journal article—primary
Educational materials:	No
Number of references:	26
Industries/occupations:	
Specific process:	
Chemical:	
Specific chemicals:	
Mixtures:	No
Audience:	Professional
Topics addressed:	E Risk assessment
	E.1 Guidelines for risk assessment or analysis
	E.1.B Systemic health effects
	E.2 Example of risk assessments
Summary:	The authors present Regetox, a two-step approach for assessing chemical health risks. The first step uses the method developed in France by the l'Institut National de Recherche et de Sécurité (INRS) to rank potential risk. The second step uses the COSHH method and EASE model established by the UK Health & Safety Executive to assess chemical risk using occupational exposure limits. The authors call Regetox a useful tool for chemical risk assessment in small- and medium-sized enterprises (MSEs).

Article ID:	**34**
Citation:	Basketter DA, Flyvholm MA, Menne T [1999]. Classification criteria for skin-sensitizing chemicals: A commentary. Contact Dermatitis *40*(4):175–82.
Resource type:	Journal article—review, meta-analysis
Educational materials:	No
Number of references:	35
Industries/occupations:	
Specific process:	
Chemical:	
Specific chemicals:	
Mixtures:	No
Audience:	Professional

Appendix A: Full Resource Citations and Summaries

Topics addressed:	A	Overview
	A.3	Investigation, intervention, and control of occupational skin exposures
	A.5	Dermal regulations and skin notations
	D	Hazard identification
	D.1	Potential health effects resulting from specific chemicals
	D.1.A	Irritant contact dermatitis
	D.1.B	Allergic contact dermatitis/sensitization
	D.1.C	Systemic toxicity
Summary:	\multicolumn{2}{l}{This article reviews the benefits and limitations of systems to classify substances with high dermal potency (either significant skin sensitizers or important contact allergens). Included are discussions of strategies used by the European Union, the World Health Organization, and the United States. Such information is necessary for proper risk assessment and management of skin sensitizers.}

Article ID:	**35**
Citation:	Basketter DA, Evans P, Gerberick GF, Kimber IA [2002]. Factors affecting thresholds in allergic contact dermatitis: Safety and regulatory considerations. Contact Dermatitis 47(1):1–6.
Resource type:	Journal article—review, meta-analysis
Educational materials:	No
Number of references:	39
Industries/occupations:	
Specific process:	
Chemical:	
Specific chemicals:	
Mixtures:	No
Audience:	Professional

Topics addressed:	A	Overview
	A.4	Skin physiology and function as barriers to chemical insults
	D	Hazard identification
	D.2	Summaries of health effects, dose-response relationships
	D.4	Other

Appendix A: Full Resource Citations and Summaries

Summary:	This article examines the nature of thresholds in allergic contact dermatitis. These thresholds vary according to whether skin exposure is transient or prolonged, open or occluded, and single or repeated, as well as the condition of the skin, the presence of inflammation, and the vehicle in which a chemical sensitizer comes into contact with the skin. Recommendations are provided for safety evaluation and dermal regulations. Allergic potencies also are provided for 38 chemicals using both guinea pig and lymph node assay data.

Article ID:	**36**	
Citation:	Baynes RE [2005]. Dermal absorption of cutting fluid mixtures. Raleigh, NC: Center for Chemical Toxicology Research and Pharmacokinetics, College of Veterinary Medicine, North Carolina State University.	
Resource type:	Technical publication/report	
Educational materials:	No	
Number of references:	38	
Industries/occupations:	Manufacturing—Machining industry	
Specific process:		
Chemical:	Coolants, heavy metals/inorganic compounds, petroleum products & lubricants, solvents, cutting fluid additives	
Specific chemicals:	Linear alkibenzene sulfonate (LAS), sulfate ricinolei acid (RA), tiazine, nickel, trichloroethylene (TCE), triethanolamine N-nitrosodiethanolamine	
Mixtures:	Yes	
Audience:	Professional	
Topics addressed:	C	Exposure characterization
	C.2	Description of factors influencing exposure conditions
	C.2.E	Uptake
	D	Hazard identification
	D.1	Potential health effects resulting from specific chemicals
	D.1.A	Irritant contact dermatitis
	D.3	Characterization protocols
	D.3.E	Measurement of skin permeation rates and reservoir effects
Summary:	This document reports results from the testing of several cutting fluid additives and contaminants to ascertain the influence of chemical mixtures on dermal disposition and	

Indexed Dermal Bibliography

Appendix A: Full Resource Citations and Summaries

cutaneous toxicity. The research examined three specific additives: linear alkibenzene sulfonate (LAS), sulfate ricinolei acid (RA), and tiazine with regard to dermal absorption, physiochemical interactions, and the effect on solvent (TCE) permeability.

Article ID:	**37**
Citation:	Bello D, Sparer J, Redlich CA, Ibrahim K, Stowe MH, Liu Y [2007]. Slow curing of aliphatic polyisocyanate paints in automotive refinishing: A potential source for skin exposure." J Occup Environ Hyg 4(6):406–11.
Resource type:	Journal article—primary
Educational materials:	No
Number of references:	26
Industries/occupations:	Manufacturing, Automotive refinishing painting
Specific process:	Autobody workers
Chemical:	Paints
Specific chemicals:	Aliphatic isocyanates, hexamethylene diisocyanate (pHDI) isophorone diisocyanate (pIPDI), HDI, MDI, TDI
Mixtures:	No
Audience:	Professional
Topics addressed:	A Overview
	A.1 Occurrence of skin exposures in workplace
	C Exposure characterization
	C.2 Description of factors influencing exposure conditions
	C.2.D Other
	C.4 Direct methods to measure exposure
	C.4.A Surfaces
Summary:	The paint of recently painted automobiles contains unbound isocyanate species which pose less risk after drying. This study investigated how long such paint takes to cure. From study results, the authors conclude that unbound isocyanates remain present up to 120 hours for typical paint formulations and for one month for others, presenting a risk to autobody workers.

Article ID:	**38**
Citation:	Bello D, Herrick CA, Smith TJ, Woskie SR, Streicher RP, Cullen MR, Liu Y, Redlich CA [2007]. Skin exposure to

isocyanates: Reasons for concern. Environ Health Perspect *115*(3):328–35.

Resource type:	Journal article—review, meta-analysis
Educational materials:	No
Number of references:	93
Industries/occupations:	Manufacturing—Chemical
Specific process:	Polyurethane production
Chemical:	Plastics and resins
Specific chemicals:	Isocyanates, methylene diisocyanate (MDI), toluene diisocyanate (TDI) polymeric hexamethylene diisocyanate (pHDI), isophorone diisocyanate (pIPDI), glues, foam insulation
Mixtures:	Yes
Audience:	Professional

Topics addressed:

A	Overview
A.1	Occurrence of skin exposures in workplace
B	Surveillance and clinical aspects
B.1	Surveillance study reporting incidences of occupational skin exposures
B.1.B	Skin exposure minor focus
C	Exposure characterization
C.1	Workplace factors associated with harmful skin exposures
C.2	Description of factors influencing exposure conditions
C.2.E	Uptake
C.4	Direct methods to measure exposure
C.4.B	Skin
D	Hazard identification
D.1	Potential health effects resulting from specific chemicals
D.1.B	Allergic contact dermatitis/sensitization
D.1.D	Other health effects
D.4	Other
F	Risk management
F.1	Exposure control strategies
F.1.D	PPE and PPE rules

Summary: This resource discusses the results of a literature review of 800 animal and human studies on isocyanate skin-exposure methods, workplace skin exposure, skin

Appendix A: Full Resource Citations and Summaries

absorption, and the role of skin exposure in isocyanate sensitization and asthma.

Article ID:	**39**
Citation:	Benford DJ, Cocker J, Sartorelli P, Schneider T, van Hemmen J, Firth JG [1999]. Dermal route in systemic exposure. Scand J Work Environ Health 25(6):511–20.
Resource type:	Journal article—review, meta-analysis
Educational materials:	No
Number of references:	31
Industries/occupations:	
Specific process:	
Chemical:	
Specific chemicals:	
Mixtures:	No
Audience:	Professional
Topics addressed:	C Exposure characterization
	C.2 Description of factors influencing exposure conditions
	C.2.E Uptake
	C.4 Direct methods to measure exposure
	C.4.A Surfaces
	C.4.B Skin
	C.4.C Biomonitoring
	C.5 Exposure modeling
	D Hazard identification
	D.3 Characterization protocols
	D.3.E Measurement of skin permeation rates and reservoir effects
	D.3.F QSARs—development, validation, and application
Summary:	This article discusses methods for measuring skin and surface contamination, biological monitoring, and estimating dermal uptake via *in vitro* and *in vivo* methods. The article also discusses how standardized components of exposure characterization can be developed, and how they can be used to support a generic approach to dermal risk assessment and allow for the development of workplace-appropriate assessment strategies.

Appendix A: Full Resource Citations and Summaries

Article ID:	**40**
Citation:	Boeniger M [2002]. Chemical protective clothing and the skin: Practical considerations. In: Anna DH, ed. Chemical Protective Clothing Series. Fairfax, VA: American Industrial Hygiene Association. 549-PC-02:1–48.
Resource type:	Book/monograph, chapter
Educational materials:	No
Number of references:	170
Industries/occupations:	
Specific process:	
Chemical:	
Specific chemicals:	
Mixtures:	No
Audience:	Professional
Topics addressed:	A Overview
	A.4 Skin physiology and function as barriers to chemical insults
	B Surveillance and clinical aspects
	B.1 Surveillance study reporting incidences of occupational skin exposures
	B.1.A Skin exposure major focus
	D Hazard identification
	D.1 Potential health effects resulting from specific chemicals
	D.1.A Irritant contact dermatitis
	F Risk management
	F.1 Exposure control strategies
	F.1.D PPE and PPE rules
Summary:	This chapter describes environmental and nonenvironmental factors affecting dermal absorption and how PPE reduces absorption. Factors affecting dermal absorption that are covered include: anatomical differences, interindividual differences, physical damage to skin and temperature, humidity, the chemical vehicle, and skin occlusion.

Article ID:	**41**
Citation:	Boeniger MF [2000]. Exposure and absorption of hazardous materials through the skin. Int J Occup Environ Health 6(2):148–50.

Appendix A: Full Resource Citations and Summaries

Resource type:	Other—commentary
Educational materials:	No
Number of references:	19
Industries/occupations:	
Specific process:	
Chemical:	
Specific chemicals:	
Mixtures:	No
Audience:	Professional

Topics addressed:		
	A	Overview
	A.5	Dermal regulations and skin notations
	C	Exposure characterization
	C.2	Description of factors influencing exposure conditions
	C.2.E	Uptake

Summary: This letter to the editor from NIOSH summarizes the history of attempts to quantify dermal permeation rates. Rates, based upon LD50 or permeation coefficients from saturated aqueous solutions, form the bases for OSHA skin notations, ACGIH threshold limit values (TLVs), OSHA permissible exposure limits (PELs), and NIOSH recommended exposure limits (RELs). As a result of differing laboratory methodologies, these limits may vary by several orders of magnitude. The letter offers recommendations for those offering additional skin exposure guidance and permeation criteria.

Article ID:	**42**
Citation:	Boeniger MF, Klingner TD [2002]. In-use testing and interpretation of chemical-resistant glove performance. Appl Occup Environ Hyg *17*(5):368–78.
Resource type:	Journal article—review, meta-analysis
Educational materials:	No
Number of references:	49
Industries/occupations:	
Specific process:	
Chemical:	
Specific chemicals:	
Mixtures:	No
Audience:	Professional
Topics addressed:	B Surveillance and clinical aspects

Appendix A: Full Resource Citations and Summaries

	B.2	Loss of workdays and impact on productivity
	C	Exposure characterization
	C.4	Direct methods to measure exposure
	C.4.B	Skin
	C.5	Exposure modeling
	F	Risk management
	F.1	Exposure control strategies
	F.1.D	PPE and PPE rules
Summary:	This article reviews methods for testing glove performance during actual use and offers a method for estimating acceptable exposure guidance criteria for evaluation of chemicals that are systemically absorbed.	

Article ID:	**43**
Citation:	Boeniger MF, Ahlers HW [2003]. Federal government regulation of occupational skin exposure in the USA. Int Arch Occup Environ Health 76(5):387–99.
Resource type:	Journal article—review, meta-analysis
Educational materials:	No
Number of references:	40
Industries/occupations:	
Specific process:	
Chemical:	
Specific chemicals:	
Mixtures:	No
Audience:	General
Topics addressed:	A Overview
	A.3 Dermal regulations and skin notations
Summary:	This paper provides an overview of federal regulations of dermal exposure. An analysis of 14 federal regulations and three agencies that regulate occupational skin exposure in the United States is presented. USEPA requires reporting of chemical health effects information which it uses to assess exposure risk. The Food and Drug Administration (FDA) regulates the labeling of cosmetics and requires safety data on new health products. OSHA regulates workplace safety and assesses compliance through field inspections. This paper evaluates how well the regulations prevent exposure and recommends measures to further protect workers from occupational skin hazards.

Appendix A: Full Resource Citations and Summaries

Article ID:	**44**
Citation:	Boman A, Maibach HI [2000]. Percutaneous absorption of organic solvents. Int J Occup Environ Health 6(2):93–95.
Resource type:	Journal article—review, meta-analysis
Educational materials:	No
Number of references:	37
Industries/occupations:	
Specific process:	
Chemical:	Solvents
Specific chemicals:	n-Butanol, toluene, 1,1,1-trichloroethane
Mixtures:	No
Audience:	Professional
Topics addressed:	C Exposure characterization
	C.2 Description of factors influencing exposure conditions
	C.2.E Uptake
	C.5 Exposure modeling
	D Hazard identification
	D.2 Summaries of health effects, dose-response relationships
Summary:	This paper discusses factors affecting percutaneous absorption of organic solvents including the pathway, toxicity, and environmental factors. Absorption rates vary considerably. Amphiphilic solvents have higher absorption rates. Nonoccluded repeated exposure results in less absorption than continuous contact. Ventilation reduces absorption.

Article ID:	**45**
Citation:	Boman A [2005]. Protective gloves for occupational use. Boca Raton, FL: CRC Press.
Resource type:	Book/monograph, whole
Educational materials:	No
Number of references:	1014
Industries/occupations:	Service—Medical
Specific process:	
Chemical:	Latex, pesticides
Specific chemicals:	
Mixtures:	No

Audience:	Professional	
Topics addressed:	A	Overview
	A.5	Dermal regulations and skin notations
	F	Risk management
	F.1	Exposure control strategies
	F.1.D	PPE and PPE rules
Summary:	This book covers quality standards and requirements related to the side effects of glove use, advanced technical standard test method results, permeation and penetration test data, medical reports on side effects, and applications in the glove selection process. It discusses protective glove use safety and use directives, regulations, and requirements in Europe and the United States, standard quality control test methods, in vivo testing with animals, and clinical diagnostic testing in patients.	

Article ID:	**46**	
Citation:	Bos PM, Brouwer DH, Stevenson H, Boogaard PJ, de Kort WL, van Hemmen JJ [1998]. Proposal for the assessment of quantitative dermal exposure limits in occupational environments, part 1. Development of a concept to derive a quantitative dermal occupational exposure limit. Occup Environ Med 55(12):795–804.	
Resource type:	Journal article—review, meta-analysis	
Educational materials:	No	
Number of references:	48	
Industries/occupations:		
Specific process:		
Chemical:		
Specific chemicals:	cyclophosphamide, 4,4-methylene dianiline (MDA)	
Mixtures:	No	
Audience:	Professional	
Topics addressed:	A	Overview
	A.5	Dermal regulations and skin notations
	C	Exposure characterization
	C.2	Description of factors influencing exposure conditions
	C.2.A	Exposure intensity/frequency/duration
	C.2.B	Exposure concentration
	C.2.C	Skin area affected

Appendix A: Full Resource Citations and Summaries

	C.2.D	Other
	C.2.E	Uptake
	D	Hazard identification
	D.1	Potential health effects resulting from specific chemicals
	D.1.A	Irritant contact dermatitis
	D.1.B	Allergic contact dermatitis/sensitization
	D.3	Characterization protocols
	D.3.E	Measurement of skin permeation rates and reservoir effects
Summary:	The authors argue that quantitative dermal occupational exposure limits (DOEL) should be developed, similar to respiratory occupational exposure limits (OELs), to replace today's qualitative "skin notation" warnings. The authors present a procedure for developing DOELs for the total dose deposited on the skin during a working shift and use their procedure to develop a DOEL for cyclophosphamide and 4,4-methylene dianiline (MDA). They conclude that the DOEL that they developed is relevant and useful, but further research is needed to show whether the procedure is applicable to other chemicals.	

Article ID:	**47**	
Citation:	CDC [2002]. Guideline for hand hygiene in health-care settings. Recommendations of the Healthcare Infection Control Practices Advisory Committee and the HICPAC/SHEA/APIC/IDSA Hand Hygiene Task Force. MMWR Recomm Rep *51*(RR16):1–45.	
Resource type:	Guideline	
Educational materials:	No	
Number of references:	423	
Industries/occupations:		
Specific process:		
Chemical:		
Specific chemicals:		
Mixtures:	No	
Audience:	Professional	
Topics addressed:	A	Overview
	A.4	Skin physiology and function as barriers to chemical insults
	C	Exposure characterization

Appendix A: Full Resource Citations and Summaries

	C.5	Exposure modeling
	F	Risk management
	F.1	Exposure control strategies
	F.1.C	Work practice/Administrative controls
	F.1.E	Skin management, barrier creams, moisturizers, cleansers, and rubs
Summary:	The *Guideline for Hand Hygiene in Health-Care Settings* provides healthcare workers with a review of data regarding handwashing, hand antisepsis, and barrier creams as well as recommendations to improve hand-hygiene practices and to reduce transmission of pathogenic microorganisms.	

Article ID:	**48**	
Citation:	Brondeau MT, Hesbert A, Beausoleil C, Schneider O [1999]. To what extent are biomonitoring data available in chemical risk assessment? Hum Exp Toxicol 18(5):322–26.	
Resource type:	Journal article—review, meta-analysis	
Educational materials:	No	
Number of references:	58	
Industries/occupations:		
Specific process:		
Chemical:	Solvents	
Specific chemicals:	Styrene; TCE; acrylonitrile; buta-1,3-diene; cyclohexane; 1,4-dichlorobenzene; hydrogen fluoride; 2-(2-methoxyethoxy)ethanol; alkanes-C10-13-chloro; benzene-C10-13-alkyl derivatives; bis(pentabromophenyl) ether; diphenyl ether, octabromo derivative	
Mixtures:	No	
Audience:	Professional	
Topics addressed:	C	Exposure characterization
	C.2	Description of factors influencing exposure conditions
	C.2.E	Uptake
	C.4	Direct methods to measure exposure
	C.4.C	Biomonitoring
Summary:	Biomonitoring information is helpful in assessing chemical exposure and would result in more accurate risk assessments. The availability of biomonitoring and metabolism animal data, skin penetration ability, and atmospheric threshold limits were examined for 12 substances:	

Appendix A: Full Resource Citations and Summaries

styrene; TCE; acrylonitrile; buta-1,3-diene; cyclohexane; 1,4-dichlorobenzene; hydrogen fluoride; 2-(2-methoxyethoxy) ethanol; alkanes-C10-13-chloro; benzene-C10-13-alkyl derivatives; bis(pentabromophenyl)ether; and diphenylether, octabromo derivative. The availability of biomonitoring data varied from widely available (for styrene and TCE) to lacking or scarce.

Article ID:	**49**
Citation:	Brouwer DH, Boeniger MF, van Hemmen J [2000]. Hand wash and manual skin wipes. Ann Occup Hyg 44(7):501–10.
Resource type:	Journal article—review, meta-analysis
Educational materials:	No
Number of references:	39
Industries/occupations:	
Specific process:	
Chemical:	
Specific chemicals:	
Mixtures:	No
Audience:	Professional
Topics addressed:	C Exposure characterization
	C.4 Direct methods to measure exposure
	C.4.B Skin
Summary:	This paper reviews both hand wash and skin wipe techniques of dermal exposure sampling for sampling efficiency. Sampling protocols hamper comparisons of study results. The authors conclude harmonization of sampling protocols will be a first step in creating a database for better understanding the influence of sampling parameters on the performance of removal techniques to assess dermal exposure.

Article ID:	**50**
Citation:	Brouwer DH, Semple S, Marquart J, Cherrie JW [2001]. A dermal model for spray painters. Part I: Subjective exposure modelling of spray paint deposition. Ann Occup Hyg 45(1):15–23.
Resource type:	Journal article—primary
Educational materials:	No

Appendix A: Full Resource Citations and Summaries

Number of references:	21
Industries/occupations:	Construction
Specific process:	Spray painting
Chemical:	
Specific chemicals:	
Mixtures:	No
Audience:	Professional

Topics addressed:		
	C	Exposure characterization
	C.1	Workplace factors associated with harmful skin exposures
	C.2	Description of factors influencing exposure conditions
	C.2.A	Exposure intensity/frequency/duration
	C.2.B	Exposure concentration
	C.2.C	Skin area affected
	C.2.E	Uptake
	C.5	Exposure modeling

Summary: Part 1 of 2. This paper presents a model based upon "a process-based, structured approach" that estimates both occupational dermal exposure and uptake of solvents, using airless spray painters as an example. Estimates are based upon spray technique, object shape, workers' individual work practices, droplet formation, and deposition. Predicted exposure showed reasonable correlation with the actual measured exposure, and the authors conclude that a structured, process-based approach has the potential to produce reliable estimates of dermal exposure, and they call for additional field studies.

Part 1 presents this "structured, subjective assessment" of dermal exposure estimation and evaluates its reliability.

Article ID:	**51**
Citation:	Brouwer DH, Aitken RJ, Oppl R, Cherrie JW [2005]. Concepts of skin protection: Considerations for the evaluation and terminology of the performance of skin protective equipment. J Occup Environ Hyg 2(9):425–34.
Resource type:	Journal article—review, meta-analysis
Educational materials:	No
Number of references:	24
Industries/occupations:	

Appendix A: Full Resource Citations and Summaries

Specific process:		
Chemical:		
Specific chemicals:		
Mixtures:	No	
Audience:	Professional	
Topics addressed:	A	Overview
	A.3	Investigation, intervention, and control of occupational skin exposures
	A.4	Skin physiology and function as barriers to chemical insults
	C	Exposure characterization
	C.2	Description of factors influencing exposure conditions
	C.2.A	Exposure intensity/frequency/duration
	C.2.B	Exposure concentration
	C.2.C	Skin area affected
	C.2.E	Uptake
	C.5	Exposure modeling
	F	Risk management
	F.1	Exposure control strategies
	F.1.D	PPE and PPE rules
Summary:	This article proposes a common dermal exposure glossary, including processes involved in transport, loading, uptake, and personal protective equipment. It presents both exposure loading and skin protective equipment models and presents performance data for skin protective equipment.	

Article ID:	**52**
Citation:	Brown JW III [2002]. Chemical hand protection. Occup Health Saf 71(2):56–68.
Resource type:	Magazine article
Educational materials:	Yes
Number of references:	0
Industries/occupations:	
Specific process:	
Chemical:	Solvents, ketones, acids, esters
Specific chemicals:	Benzene
Mixtures:	Yes

Appendix A: Full Resource Citations and Summaries

Audience:	General	
Topics addressed:	E	Risk management
	E.3	"Best practices"/guidelines/recommendations
	E.3.D	PPE and PPE rules
	E.3.E	Skin management, barrier creams, moisturizers, cleansers, and rubs
Summary:	This article summarizes the merits of various work gloves including natural rubber, nitrile, neoprene, hypalon, butyl, viton, and ethylene vinyl alcohol (EVOH). This article also discusses selection criteria including chemical resistance, finish and lining, and glove comfort and dexterity. This article also briefly discusses skin conditions and PPE training.	

Article ID:	**53**	
Citation:	Brown T [2004]. Strategies for prevention: Occupational contact dermatitis. Occup Med (Lond) 54(7):450–57.	
Resource type:	Journal article—review, meta-analysis	
Educational materials:	No	
Number of references:	109	
Industries/occupations:		
Specific process:		
Chemical:		
Specific chemicals:		
Mixtures:	No	
Audience:	Professional	
Topics addressed:	B	Surveillance and clinical aspects
	B.1	Surveillance study reporting incidences of occupational skin exposures
	B.1.A	Skin exposure major focus
	B.2	Loss of workdays and impact on productivity
	B.3	Surveillance study protocols/procedures for gathering data
	F	Risk management
	F.1	Exposure control strategies
	F.1.A	Substitution
	F.1.B	Engineering controls
	F.1.D	PPE and PPE rules

Appendix A: Full Resource Citations and Summaries

	F.1.E	Skin management, barrier creams, moisturizers, cleansers, and rubs
	F.2	Protocols for risk management
	F.2.B	Development of approach to achieve exposure reduction goal
Summary:		This paper presents strategies for preventing occupational contact dermatitis (OCD) including elimination/substitution, engineered/technical controls, personal protective equipment (PPE), identifying susceptible individuals, education, training and surveillance.

Article ID:	**54**	
Citation:	Bureau of Labor Statistics (BLS) [2005]. BLS industry illness and injury data. [www.bls.gov/iif/oshsum.htm].	
Resource type:	Web page	
Educational materials:	No	
Number of references:		
Industries/occupations:	Agricultural, Beauty/Cosmetology, Cleaning/Janitorial/Maid, Construction, Forestry/Fisheries, Manufacturing—Chemical, Manufacturing, Medical Services, Mining, Service—Food, Service—Medical, Service—Transportation/Communications/Utility, Other	
Specific process:		
Chemical:		
Specific chemicals:		
Mixtures:	No	
Audience:	Professional	
Topics addressed:	B	Surveillance and clinical aspects
	B.1	Surveillance study reporting incidences of occupational skin exposures
	B.1.A	Skin exposure major focus
	B.1.B	Skin exposure minor focus
	B.2	Loss of workdays and impact on productivity
	B.3	Surveillance study protocols/procedures for gathering data
Summary:	The BLS is an independent national statistical agency that collects, processes, analyzes, and disseminates essential statistical data to the American public, the U.S. Congress, other Federal agencies, state and local governments, business, and labor. The BLS also serves as a statistical resource to the Department of Labor. The Injuries,	

Illnesses, and Fatalities Program provides data on illnesses and injuries on the job and data on worker fatalities, summarized by year. The data are presented in different forms, including illness rates for different industries by type of illness. Skin diseases and skin disorders are one of the types of illnesses reported.

Article ID:	**55**
Citation:	Burnett CA, Lushniak BD, McCarthy W, Kaufman J [1998]. Occupational dermatitis causing days away from work in U.S. private industry, 1993. Am J Ind Med 34(6):568–73.
Resource type:	Journal article—primary
Educational materials:	No
Number of references:	15
Industries/occupations:	Agricultural, Cleaning/Janitorial/Maid, Construction, Manufacturing, Service—Food, Service—Medical, Transportation/Communications/Utility—Groundskeepers, Gardeners, Mechanics, Printing Press Operators, Repairmen
Specific process:	Provides risk data from many occupational groups.
Chemical:	General—overview, cleaning agents, petroleum products & lubricants, plastics and resins, solvents
Specific chemicals:	Calcium hydroxide
	Provides risk data for several chemical classes
Mixtures:	No
Audience:	Professional
Topics addressed:	B Surveillance and clinical aspects
	B.1 Surveillance study reporting incidences of occupational skin exposures
	B.1.A Skin exposure major focus
	B.2 Loss of workdays and impact on productivity
	B.3 Surveillance study protocols/procedures for gathering data
Summary:	The authors examined the 8,835 cases of dermal exposure in the 1993 Annual Survey of Occupational Injuries and Illnesses from the BLS. The article presents considerable surveillance data including rates of occupational dermatitis, identifies the service sector with the greatest number of cases, the sectors with the highest exposure rates, and the chemicals causing the greatest number of exposures.

Appendix A: Full Resource Citations and Summaries

Article ID:	56
Citation:	Byrne MA [2000]. Suction methods for assessing contamination on surfaces. Ann Occup Hyg *44*(7):523–28.
Resource type:	Journal article—review, meta-analysis
Educational materials:	No
Number of references:	21
Industries/occupations:	
Specific process:	
Chemical:	
Specific chemicals:	
Mixtures:	No
Audience:	Professional
Topics addressed:	C Exposure characterization
	C.4 Direct methods to measure exposure
	C.4.A Surfaces
	C.5 Exposure modeling
Summary:	This paper reviews commonly employed sampling techniques for occupational surfaces reported in the literature, removal efficiency, and applicability to dermal exposure assessment.

Article ID:	57
Citation:	California Department of Industrial Relations (CADIR), Division of Labor Statistics and Research [2003]. [www.dir.ca.gov/DLSR/Injuries/2003/Menu.htm].
Resource type:	Web site
Educational materials:	No
Number of references:	
Industries/occupations:	general—overview
Specific process:	
Chemical:	
Specific chemicals:	
Mixtures:	No
Audience:	Professional
Topics addressed:	B Surveillance and clinical aspects
	B.1 Surveillance study reporting incidences of occupational skin exposures
	B.1.A Skin exposure major focus

	B.1.B	Skin exposure minor focus
	B.2	Loss of workdays and impact on productivity
Summary:	The California Division of Labor Statistics and Research collects, compiles, and presents statistics and research relating to the condition of labor in the state. Data presented on their site include incidence rates and numbers of nonfatal occupational illnesses by industry sector and category of illness and the numbers of nonfatal occupational illnesses by selected industries and category of illness.	

Article ID:	58	
Citation:	Canadian Centre for Occupational Health and Safety (CCOHS) [2005]. [www.ccohs.ca/].	
Resource type:	Web site	
Educational materials:	No	
Number of references:		
Industries/occupations:	General—overview	
Specific process:	Multiple occupations	
Chemical:	General—overview, corrosives, heavy metals/inorganic compounds, latex, particulates, petroleum products & lubricants, solvents	
Specific chemicals:	Multiple chemicals	
Mixtures:	No	
Audience:	General	
Topics addressed:	A	Overview
	A.1	Occurrence of skin exposures in the workplace
	A.2	Health hazards resulting from skin exposure to chemicals
	B	Exposure characterization
	B.1	Job/tasks, industries/processes, or chemicals associated with skin exposures
	B.2	Factors that influence exposure conditions
	B.2.B	Exposure controls
	C	Hazard identification
	C.3	Protocols/checklists to identify skin hazards in the workplace
	E	Risk management
	E.3	"Best practices"/guidelines/recommendations
	E.3.B	Engineering controls

Appendix A: Full Resource Citations and Summaries

E.3.C	Work practice/administration controls
E.3.D	PPE and PPE rules
E.3.E	Skin management, barrier creams, moisturizers, cleansers, and rubs

Summary: The Canadian Centre for Occupational Health and Safety is a Canadian federal government agency that promotes workplace health and safety by providing resources and programs on different health and safety topics. Their Web site contains general information, articles, news releases, products, and services related to occupational safety and health. Dermal exposure-related information can be found throughout the site. Good sources include:

- OSH Answers—This searchable page contains information on different topics on occupational health and safety. Information can be searched based on hazards present; occupations and workplaces; and diseases, disorders, and injuries, to name a few. OSH Answers contains hazard and prevention-related information for a variety of chemicals and chemical classes that are known skin hazards, including allergic contact dermatitis.

- Dermatitis, Allergic Contact Web page—This Web page covers occupations at risk, recognition, treatment, and preventive measures associated with allergic contact dermatitis.

- Dermatitis, Irritant Contact Web page—This Web page covers occupations at risk, recognition, treatment, and preventive measures associated with irritant contact dermatitis.

- WHMIS (Workplace Hazardous Materials Information System) label contains health hazard information. These labels are required by law. They use classifications to group chemicals with similar properties or hazards. Class E-corrosive material is a classification for compounds that can cause burns to eyes, skin, or respiratory system.

Article ID:	**59**
Citation:	The Center to Protect Workers' Rights (CPWR) [2006]. [www.cpwr.com/indexstart.html].
Resource type:	Web site
Educational materials:	Yes

Appendix A: Full Resource Citations and Summaries

Number of references:		
Industries/occupations:	Construction	
Specific process:		
Chemical:	General—overview, corrosives, solvents	
Specific chemicals:		
Mixtures:	No	
Audience:	General	
Topics addressed:	A	Overview
	A.1	Occurrence of skin exposures in the workplace
	C	Hazard identification
	C.2	Tables/charts/lists of hazards for specific chemicals
	E	Risk management
	E.1	Overview of skin exposure control options
	E.3	"Best practices"/guidelines/recommendations
	E.3.C	Work practice/administration controls
	E.3.D	PPE and PPE rules
	E.3.E	Skin management, barrier creams, moisturizers, cleansers, and rubs
	E.4	Guidelines/recommendations for postexposure skin decontamination

Summary: CPWR is a nonprofit organization created by the Building and Construction Trades Department of the AFL-CIO. They provide applied research, training, and service to the construction industry. CPWR developed and maintains eLCOSH (for more information see eLCOSH, ID 63), which provides online construction-related health and safety information in English, Spanish, and other languages. The CPWR Web site also contains updates on conferences, publications, and news events associated with construction health and safety. Dermal exposure-related resources available on this Web site include:

- Hazard Alerts: *Skin Problems in Construction, Beryllium, Solvents, and Lead*
- *The Construction Chart Book*, which contains a chapter titled "Nonfatal Skin Diseases and Disorders in Construction."
- *Save Your Skin: Tips on Preventing Skin Problems* (brochure)
- The Construction Solutions Database, currently under development, will organize hazards by tasks and present ways to control those hazards.

Indexed Dermal Bibliography **133**

Appendix A: Full Resource Citations and Summaries

Article ID:	60
Citation:	CPWR [2000]. A safety and health practitioner's guide to skin protection. [www.cdc.gov/elcosh/docs/d0400/d000458/d000458.html].
Resource type:	Brochure, pamphlet
Educational materials:	Yes
Number of references:	
Industries/occupations:	Construction
Specific process:	Bricklayer, Carpenter, Masons, Hod Carrier, Plasterer, Terrazzo Worker, Tile Setter
Chemical:	Corrosives, hand cleansers, heavy metals/inorganic compounds, plastics and resins
Specific chemicals:	Portland cement, CrVI
Mixtures:	No
Audience:	Professional
Topics addressed:	A Overview
	A.1 Occurrence of skin exposures in the workplace
	A.2 Health hazards resulting from skin exposure to chemicals
	A.4 Skin physiology and function as barriers to chemical insults
	B Surveillance and clinical aspects
	B.1 Surveillance study reporting incidences of occupational skin exposures
	B.1.A Skin exposure major focus
	B.2 Loss of workdays and impact on productivity
	B.3 Surveillance study protocols/procedures for gathering data
	B.4 Clinical protocols for recognition of skin exposure health effects
	C Exposure characterization
	C.1 Workplace factors associated with harmful skin exposures
	C.2 Description of factors influencing exposure conditions
	C.2.A Exposure intensity/frequency/duration
	C.2.B Exposure concentration
	C.2.E Uptake
	C.3 Checklists/questionnaires to quantify skin exposure incidences

Appendix A: Full Resource Citations and Summaries

C.4	Direct methods to measure exposure
C.4.A	Surfaces
D	Hazard identification
D.1	Potential health effects resulting from specific chemicals
D.1.A	Irritant contact dermatitis
D.1.B	Allergic contact dermatitis/sensitization
D.1.D	Other health effects
E	Risk assessment
E.1	Guidelines for risk assessment or analysis
E.1.A	Localized health effects
F	Risk management
F.1	Exposure control strategies
F.1.A	Substitution
F.1.B	Engineering controls
F.1.C	Work practice/Administrative controls
F.1.D	PPE and PPE rules
F.1.E	Skin management, barrier creams, moisturizers, cleansers, and rubs
F.2	Protocols for risk management
F.2.C	Evaluation to demonstrate program/intervention effectiveness

Summary: This comprehensive guide was developed by the CPWR Consortium on Preventing Contact Dermatitis. Developed for safety and health practitioners, it covers dermal hazards associated with Portland cement work. It provides a detailed description of how to recognize, evaluate, and control dermal hazards. In addition to covering Portland cement, it also provides a thorough description of skin physiology and presents a model of skin disease within the context of occupational exposures to caustic chemicals and sensitizing agents. At the end, it provides a list of recommended resources, a best practices checklist, a skin symptoms questionnaire for workers, and a worker safety pamphlet. This brochure is available for download on the eLCOSH Web site (see Article ID 63).

This guide contains the following chapters:

Ch. 1	Recognizing Skin Problems
Ch. 2	A New Model of Skin Disease
Ch. 3	Worksite Exposures

Appendix A: Full Resource Citations and Summaries

Ch. 4	The Role of pH
Ch. 5	Product Modification
Ch. 6	Best Protective Practices
Ch. 7	Resources
Ch. 8	Evaluating Your Success

Article ID:	**61**	
Citation:	CPWR [2001]. Physician's alert for occupational contact dermatitis among construction workers. [www.cdc.gov/elcosh/docs/d0200/d000281/d000281.pdf].	
Resource type:	Brochure, pamphlet	
Educational materials:	Yes	
Number of references:		
Industries/occupations:	General—overview, Construction	
Specific process:		
Chemical:	Corrosives, hand cleansers	
Specific chemicals:	Portland cement, CrVI, lanolin	
Mixtures:	No	
Audience:	Professional	
Topics addressed:	A	Overview
	A.1	Occurrence of skin exposures in workplace
	A.2	Health hazards resulting from skin exposure to chemicals
	B	Surveillance and clinical aspects
	B.4	Clinical protocols for recognition of skin exposure health effects
	F	Risk management
	F.1	Exposure control strategies
	F.1.E	Skin management, barrier creams, moisturizers, cleansers, and rubs
Summary:	Developed by the CPWR, this physician's alert pamphlet was designed as information construction workers could bring with them on visits to a physician's office for skin-related disorders. The pamphlet contains a table of skin disorders, possible work-related causes, diagnostic aids, and intervention and treatment options.	
	This pamphlet can be found online at the eLCOSH Web site (http://www.cdc.gov/elcosh/index.html).	

Appendix A: Full Resource Citations and Summaries

Article ID:	**62**
Citation:	CPWR [2005]. An employer's guide to skin protection. [www.cdc.gov/elcosh/docs/d0400/d000457/d000457.html].
Resource type:	Brochure, pamphlet
Educational materials:	Yes
Number of references:	25
Industries/occupations:	Construction
Specific process:	Bricklayers, Carpenters, Masons, Hod Carrier, Plasterer, Terrazzo Worker, Tile Setters
Chemical:	Corrosives
Specific chemicals:	Portland cement, CrVI
Mixtures:	No
Audience:	General

Topics addressed:

A	Overview
A.1	Occurrence of skin exposures in the workplace
A.2	Health hazards resulting from skin exposure to chemicals
B	Exposure characterization
B.1	Job/tasks, industries/processes, or chemicals associated with skin exposures
B.2	Factors that influence exposure conditions
B.2.A	Exposure intensity/frequency
B.2.B	Exposure controls
E	Risk management
E.1	Overview of skin exposure control options
E.3	"Best practices"/guidelines/recommendations
E.3.C	Work practice/administration controls
E.3.D	PPE and PPE rules
E.4	Guidelines/recommendations for postexposure skin decontamination

Summary: This handbook for employers in the cement product industry (concrete, mortar, plaster, grout, stucco, and terrazzo) covers issues associated with dermal exposure identification, evaluation, control, and prevention. It offers recommendations to prevent employee skin problems.

Article ID:	**63**
Citation:	CPWR [2005]. Electronic library of construction occupational safety and health (eLCOSH). [www.cdc.gov/elcosh/].

Appendix A: Full Resource Citations and Summaries

Resource type:	Web site
Educational materials:	Yes
Number of references:	
Industries/occupations:	Construction
Specific process:	
Chemical:	Corrosives, hand cleansers, heavy metals/inorganic compounds, soaps and detergents
Specific chemicals:	Portland cement, CrVI
Mixtures:	Yes
Audience:	General

Topics addressed:

A	Overview
A.1	Occurrence of skin exposures in the workplace
A.2	Health hazards resulting from skin exposure to chemicals
B	Exposure characterization
B.1	Job/tasks, industries/processes, or chemicals associated with skin exposures
B.2	Factors that influence exposure conditions
B.2.B	Exposure controls
B.3	Protocols/checklists to characterize exposure to skin hazards
C	Hazard identification
C.3	Protocols/checklists to identify skin hazards in the workplace
D	Risk assessment
D.1	Protocols/checklists to identify exposure risk
E	Risk management
E.3	"Best practices"/guidelines/recommendations
E.3.B	Engineering controls
E.3.C	Work practice/administration controls
E.3.D	PPE and PPE rules
E.3.E	Skin management, barrier creams, moisturizers, cleansers, and rubs
E.4	Guidelines/recommendations for postexposure skin decontamination

Summary: The eLCOSH is intended to provide accurate, user-friendly information about safety and health for construction workers from different sources. The eLCOSH was developed by the CPWR through funding from NIOSH and is maintained by CPWR. Information on the Web

Appendix A: Full Resource Citations and Summaries

site can be located by hazard, trade, job site, and other categories. Downloadable resources on this site related to dermal exposure to chemicals include the following:

- *Save your Skin: Tips on Preventing Skin Problems*—a general information brochure for workers.

- *Chemical Glove Selection*—a document produced by the University of Delaware Cooperative Extension on glove selection in agricultural settings.

- *Physician's Alert: Skin Conditions*—a brochure produced by the CPWR for workers to bring to their physicians office.

- *An Employer's Guide To Skin Protection*—a comprehensive document for employers covering a variety of issues associated with the evaluation, control, and prevention of dermal exposure to cement products, CrVI, and worksite cleansers.

- *A Safety & Health Practitioner's Guide to Skin Protection*—a comprehensive document on dermal exposure developed for the person responsible for protecting the safety and health of workers using Portland cement products. Similar to the document produced for employers, this document goes into more depth and includes a worker safety pamphlet, a best practices checklist, and a symptoms questionnaire.

Article ID:	64	
Citation:	CDC [2006]. NASD: National Ag Safety Database. [www.cdc.gov/nasd/].	
Resource type:	Web site	
Educational materials:	Yes	
Number of references:		
Industries/occupations:	Agricultural	
Specific process:		
Chemical:	Hand cleansers, pesticides	
Specific chemicals:		
Mixtures:	No	
Audience:	General	
Topics addressed:	E	Risk management
	E.1	Overview of skin exposure control options
	E.3	"Best practices"/guidelines/recommendations

Appendix A: Full Resource Citations and Summaries

E.3.B	Engineering controls
E.3.C	Work practice/administration controls
E.3.D	PPE and PPE rules
E.3.E	Skin management, barrier creams, moisturizers, cleansers, and rubs
E.4	Guidelines/recommendations for postexposure skin decontamination

Summary: The National Ag Safety Database (NASD), developed through funding from NIOSH, is designed to provide a national information resource on agricultural safety and health issues, with the purpose of reducing agricultural work-related illnesses and injuries. It contains a variety of agricultural technical bulletins by topic, as well as posters and videos for training. They also have Spanish language material. The materials on this Web site are produced by other sources, mostly state and university agriculture extension offices. Dermal exposure bulletins include: *Skin Irritants*, *Pesticide-Contaminated Clothing Laundering*, and general information in *Pesticide Exposure*. Under the topic "mixing, loading, and application" of chemicals/pesticides, there are a number of bulletins on the selection, proper use, cleaning, and handling of PPE. Also available are bulletins on other control measures and training videotapes on pesticide safety.

Article ID:	65
Citation:	Chang H [2005]. Total body burden arising from a week's repeated dermal exposure to N,N-dimethylformamide (DMF). Occup Environ Med 62(3):151–56.
Resource type:	Journal article—primary
Educational materials:	No
Number of references:	39
Industries/occupations:	
Specific process:	
Chemical:	Solvents
Specific chemicals:	N,N-dimethylformamide (DMF)
Mixtures:	No
Audience:	Professional
Topics addressed:	C Exposure characterization
	C.2 Description of factors influencing exposure conditions
	C.2.A Exposure intensity/frequency/duration

Appendix A: Full Resource Citations and Summaries

C.2.B	Exposure concentration
C.2.C	Skin area affected
C.2.E	Uptake
C.4	Direct methods to measure exposure
C.4.B	Skin
C.4.C	Biomonitoring
D	Hazard identification
D.2	Summaries of health effects, dose-response relationships

Summary: This paper presents the results of a study designed to estimate the contribution of skin absorption to total body burden of N,N-dimethylformamide (DMF). The study monitored 45 industrial workers and 20 nonDMF-exposed subjects for DMF exposure via respiratory and dermal routes. The control group showed no detectable exposure. The study concluded that dermal exposure to DMF can result in significant accumulation of DMF.

Article ID:	66
Citation:	OPPTS [1996]. Applications International Corporation. Occupational dermal exposure assessment: a review of methodologies and field data: final report. Washington, DC: USEPA, Economics, Exposure and Technology Division, OPPTS.EPA 600-R-96-000 9-30-1996.
Resource type:	Technical publication/report
Educational materials:	No
Number of references:	108
Industries/occupations:	
Specific process:	
Chemical:	
Specific chemicals:	
Mixtures:	No
Audience:	Professional
Topics addressed:	

C	Exposure characterization
C.4	Direct methods to measure exposure
C.4.A	Surfaces
C.4.B	Skin
C.5	Exposure modeling
D	Hazard identification

Appendix A: Full Resource Citations and Summaries

	D.2	Summaries of health effects, dose-response relationships
Summary:		This paper summarizes a literature review on dermal exposure assessment and sampling methods. It includes a review of monitoring data on dermal exposure and identifies other methods used for predicting dermal exposure when monitoring data is not available. It evaluates the method for predicting dermal exposure developed by the Chemical Engineering Branch (CEB) of the USEPA OPPTS. The CEB method is evaluated under various scenarios and the authors revise or identify additional values and input parameters (e.g., quantity remained on skin, skin surface area). This review also makes recommendations to improve the CEB method.

Article ID:	67	
Citation:	Cherrie JW, Brouwer DH, Roff M, Vermeulen R, Kromhout H [2000]. Use of qualitative and quantitative fluorescence techniques to assess dermal exposure. Ann Occup Hyg 44(7):519–22.	
Resource type:	Journal article—review, meta-analysis	
Educational materials:	No	
Number of references:	16	
Industries/occupations:		
Specific process:		
Chemical:		
Specific chemicals:		
Mixtures:	No	
Audience:	Professional	
Topics addressed:	C	Exposure characterization
	C.4	Direct methods to measure exposure
	C.4.B	Skin
Summary:		This paper reviews the literature on both quantitative and qualitative methods of dermal exposure using fluorescent tracers to estimate chemical uptake through the skin.

Article ID:	68
Citation:	Cherry N, Meyer JD, Adisesh A, Brooke R, Owen-Smith V, Swales C, Beck MH [2000]. Surveillance of occupational skin disease: EPIDERM and OPRA. Br J Dermatol 142(6).

Appendix A: Full Resource Citations and Summaries

Resource type:	Journal article—primary
Educational materials:	No
Number of references:	19
Industries/occupations:	General—overview
Specific process:	Presents data from several large occupational groups, such as chemical operatives, metal assemblers, machine tool operatives, glass manufacturers, and printers.
Chemical:	General—overview
Specific chemicals:	Presents data by several chemical agents, such as rubber, soaps, wet work, petroleum, and nickel.
Mixtures:	No
Audience:	Professional
Topics addressed:	A Overview
	A.1 Occurrence of skin exposures in workplace
	B Surveillance and clinical aspects
	B.1 Surveillance study reporting incidences of occupational skin exposures
	B.1.A Skin exposure major focus
Summary:	This paper presents summary surveillance data, collected in the U.K., from the Occupational Physicians Reporting Activity (OPRA) and its predecessor, EPIDERM. OPRA is a voluntary surveillance mechanism that has collected occupational skin disease data from dermatologists and occupational physicians since 1993. Incidences by gender, age, occupational group, and chemical group are presented.

Article ID:	**69**
Citation:	Chew AL, Maibach HI [2003]. Occupational issues of irritant contact dermatitis. Int Arch Occup Environ Health 76(5):339–46.
Resource type:	Journal article—review, meta-analysis
Educational materials:	No
Number of references:	62
Industries/occupations:	Agricultural, Beauty/Cosmetology, Cleaning/Janitorial/Maid, Construction, Manufacturing, Service—Food, Service—Medical
Specific process:	Includes table of job categories and associated irritants
Chemical:	Cleaning agents, corrosives, pesticides, soaps and detergents, solvents
Specific chemicals:	

Indexed Dermal Bibliography   **143**

Appendix A: Full Resource Citations and Summaries

Mixtures:	No	
Audience:	Professional	
Topics addressed:	A	Overview
	A.2	Health hazards resulting from skin exposure to chemicals
	B	Surveillance and clinical aspects
	B.1	Surveillance study reporting incidences of occupational skin exposures
	B.1.A	Skin exposure major focus
	B.2	Loss of workdays and impact on productivity
	B.4	Clinical protocols for recognition of skin exposure health effects
	C	Exposure characterization
	C.4	Direct methods to measure exposure
	C.4.B	Skin
	D	Hazard identification
	D.1	Potential health effects resulting from specific chemicals
	D.1.A	Irritant contact dermatitis
	F	Risk management
	F.1	Exposure control strategies
	F.1.D	PPE and PPE rules
	F.1.E	Skin management, barrier creams, moisturizers, cleansers, and rubs
Summary:	This paper reviews the various types of occupational irritant contact dermatitis along with epidemiological data, risk factors, pathophysiology, diagnosis, and management of irritant contact dermatitis.	

Article ID:	**70**
Citation:	CLI [1999]. A guide to dermal exposure reduction. Des Plaines, IL: Colormetric Laboratories, Incorporated.
Resource type:	Other—Guideline from private lab
Educational materials:	No
Number of references:	0
Industries/occupations:	
Specific process:	
Chemical:	
Specific chemicals:	

Appendix A: Full Resource Citations and Summaries

Mixtures:	No	
Audience:	Professional	
Topics addressed:	F	Risk management
	F.1	Exposure control strategies
	F.1.D	PPE and PPE rules
	F.2	Protocols for risk management
	F.2.B	Development of approach to achieve exposure reduction goal
Summary:	This pamphlet briefly outlines the benefit and contents of a dermal exposure reduction program. This pamphlet is available for download on their Web site (www.clilabs.com).	

Article ID:	**71**	
Citation:	Colormetric Laboratories, Incorporated (CLI), 2005. [www.clilabs.com/].	
Resource type:	Web site	
Educational materials:	No	
Number of references:		
Industries/occupations:	General—overview	
Specific process:		
Chemical:	General—overview, plastics and resins	
Specific chemicals:	isocyanates	
Mixtures:	No	
Audience:	Professional	
Topics addressed:	C	Exposure characterization
	C.4	Direct methods to measure exposure
	C.4.A	Surfaces
	C.4.B	Skin
	C.4.C	Biomonitoring
	E	Risk assessment
	E.1	Guidelines for risk assessment or analysis
	E.1.A	Localized health effects
	E.1.B	Systemic health effects
	F	Risk management
	F.1	Exposure control strategies
	F.1.A	Substitution
	F.1.B	Engineering controls

Indexed Dermal Bibliography **145**

Appendix A: Full Resource Citations and Summaries

F.1.C	Work practice/Administrative controls
F.1.D	PPE and PPE rules
F.1.E	Skin management, barrier creams, moisturizers, cleansers, and rubs

Summary: Colormetric Laboratories, Inc., provides biological monitoring analytical services and direct reading detection systems for evaluating surface and skin contamination. They produce the following detection systems:

- SWYPES™ detectors to determine where and when the skin exposures occurred that contributed to biological exposures.
- Permea-Tec™ Sensors which are breakthrough indicators for chemical protective gloves and can be used to determine glove life expectancy.
- D-TAM™ Safe Solvent for removing lipophilic compounds from skin.
- D-TAM™ Skin Cleansers, which are formulated to remove water-insoluble contaminants from the skin.

Article ID:	72
Citation:	Crassweller I [1999]. Helping hands: Skin care for the hands. Occup Hazards 61(8):58.
Resource type:	Magazine article
Educational materials:	No
Number of references:	0
Industries/occupations:	
Specific process:	
Chemical:	Cleaning agents
Specific chemicals:	
Mixtures:	No
Audience:	General
Topics addressed:	E Risk management
	E.3 "Best practices"/guidelines/recommendations
	E.3.E Skin management, barrier creams, moisturizers, cleansers, and rubs
	E.4 Guidelines/recommendations for postexposure skin decontamination
Summary:	Industrial workers exposed to harsh chemicals or who perform frequent hand-washing are susceptible to occupational skin diseases. This paper provides an

overview of skin care and outlines the correct method of cleaning hands.

Article ID:	**73**
Citation:	Das R, Steege A, Baron S, Beckman J, Harrison R [2001]. Pesticide-related illness among migrant farm workers in the United States. Int J Occup Environ Health 7(4):303–12.
Resource type:	Journal article—review, meta-analysis
Educational materials:	No
Number of references:	79
Industries/occupations:	Agricultural
Specific process:	Migrant farm workers
Chemical:	Pesticides
Specific chemicals:	Organophosphates, carbamates, inorganic compounds, pyrethroids
Mixtures:	No
Audience:	Professional
Topics addressed:	A Overview
	A.2 Health hazards resulting from skin exposure to chemicals
	B Surveillance and clinical aspects
	B.1 Surveillance study reporting incidences of occupational skin exposures
	B.1.A Skin exposure major focus
	F Risk management
	F.1 Exposure control strategies
	F.1.A Substitution
	F.1.C Work practice/Administrative controls
Summary:	This paper reviews a few pesticide categories (organophosphates, carbamates, inorganic compounds, and pyrethroids) that account for over half of all acute occupational illness cases among migrant farm workers in U.S. Most cases are caused by dermal exposures. Pesticide risk assessment should be based on acute toxicity, chronic toxicity, carcinogenic potency, volume applied, and magnitude of worker poisonings. Also discussed is the hierarchy of control measures, with a focus on substitution, establishing effective protections, enforcement, and education. This paper also contains a considerable amount of surveillance data.

Appendix A: Full Resource Citations and Summaries

Article ID:	**74**
Citation:	Day GA, Stefaniak AB, Weston A, and Tinkle SS [2006]. Beryllium exposure: dermal and immunological considerations. Int Arch Occup Environ Health 79(2):161–64.
Resource type:	Journal article—review, meta-analysis
Educational materials:	No
Number of references:	38
Industries/occupations:	
Specific process:	
Chemical:	
Specific chemicals:	Beryllium
Mixtures:	No
Audience:	Professional
Topics addressed:	B Surveillance and clinical aspects
	B.1 Surveillance study reporting incidences of occupational skin exposures
	B.1.A Skin exposure major focus
	D Hazard identification
	D.1 Potential health effects resulting from specific chemicals
	D.1.B Allergic contact dermatitis/sensitization
	D.1.C Systemic toxicity
Summary:	This paper assesses the state of existing knowledge concerning skin exposure to beryllium. It concludes that the reduction in inhalation exposure to beryllium has not reduced beryllium sensitization or chronic beryllium diseases (CBD), suggesting that unchecked skin exposure is causing continued prevalence.

Article ID:	**75**
Citation:	Del Rosso J [2001]. Protecting the hand-skin barrier in the workplace. Occup Health Saf 70(9):116–20.
Resource type:	Magazine article
Educational materials:	No
Number of references:	3
Industries/occupations:	
Specific process:	
Chemical:	

Appendix A: Full Resource Citations and Summaries

Specific chemicals:		
Mixtures:	No	
Audience:	General	
Topics addressed:	A	Overview
	A.2	Health hazards resulting from skin exposure to chemicals
	E	Risk management
	E.3	"Best practices"/guidelines/recommendations
	E.3.D	PPE and PPE rules
	E.3.E	Skin management, barrier creams, moisturizers, cleansers, and rubs
Summary:	This article recommends steps for proper skin care, including gloves and barrier creams.	

Article ID:	**76**	
Citation:	Diepgen TL, Coenraads PJ [1995]. What can we learn from epidemiological studies on irritant contact dermatitis? Curr Probl Dermatol *23*:18–27.	
Resource type:	Journal article—review, meta-analysis	
Educational materials:	No	
Number of references:	18	
Industries/occupations:	General—overview, Beauty/Cosmetology, Construction, Manufacturing—Other, Medical Services	
Specific process:	Electroplaters, Metalworkers, Bricklayers	
Chemical:		
Specific chemicals:		
Mixtures:	No	
Audience:	Professional	
Topics addressed:	B	Surveillance and clinical aspects
	B.1	Surveillance study reporting incidences of occupational skin exposures
	B.1.A	Skin exposure major focus
	B.3	Surveillance study protocols/procedures for gathering data
	C	Exposure characterization
	C.1	Workplace factors associated with harmful skin exposures

Indexed Dermal Bibliography **149**

Appendix A: Full Resource Citations and Summaries

Summary:	This paper presents irritant contact dermatitis surveillance data and describes how age, gender, and race affect incidence and prevalence of irritant contact dermatitis.

Article ID:	**77**	
Citation:	Diepgen TL [1996]. Epidemiological studies on the prevention of occupational contact dermatitis. Curr Probl Dermatol 25:1–9.	
Resource type:	Journal article—review, meta-analysis	
Educational materials:	No	
Number of references:	13	
Industries/occupations:	Beauty/Cosmetology	
Specific process:	Provides incidence among several occupational categories.	
Chemical:		
Specific chemicals:		
Mixtures:	No	
Audience:	Professional	
Topics addressed:	B	Surveillance and clinical aspects
	B.1	Surveillance study reporting incidences of occupational skin exposures
	B.1.A	Skin exposure major focus
	D	Hazard identification
	D.1	Potential health effects resulting from specific chemicals
	D.1.B	Allergic contact dermatitis/sensitization
	D.2	Summaries of health effects, dose-response relationships
Summary:	This paper answers the following questions about occupational contact dermatitis (OCD):	
	1. What is its public health importance?	
	2. How large are prevalence and incidence?	
	3. What industries are associated with higher risk?	
	4. What are the occupational exposures that cause it?	
	5. Who is at risk?	
	6. What is the prognosis for patients?	
	7. What preventative measures and interventions are effective?	

Appendix A: Full Resource Citations and Summaries

Article ID:	**78**
Citation:	Diepgen TL, Coenraads PJ [1999]. The epidemiology of occupational contact dermatitis. Int Arch Occup Environ Health 72(8):496–506.
Resource type:	Journal article—review, meta-analysis
Educational materials:	No
Number of references:	75
Industries/occupations:	General—overview
Specific process:	Discussion of frequency of occupational contact dermatitis (OCD) among certain occupational categories.
Chemical:	
Specific chemicals:	Dichromate
Mixtures:	No
Audience:	Professional

Topics addressed:

A	Overview
A.2	Health hazards resulting from skin exposure to chemicals
B	Surveillance and clinical aspects
B.1	Surveillance study reporting incidences of occupational skin exposures
B.1.A	Skin exposure major focus
D	Hazard identification
D.1	Potential health effects resulting from specific chemicals
D.1.A	Irritant contact dermatitis
D.1.B	Allergic contact dermatitis/sensitization
D.3	Characterization protocols
D.3.B	Irritation potential
D.3.C	Sensitization potential
F	Risk management
F.1	Exposure control strategies
F.1.D	PPE and PPE rules
F.1.E	Skin management, barrier creams, moisturizers, cleansers, and rubs

Summary: This review article discusses the lack of epidemiologic data on occupational contact dermatitis (OCD). It also discusses case ascertainment and bias, distribution of occupational allergic and irritant contact dermatitis, the interrelationship between exogenous (allergens, irritants) and endogenous factors, the prognosis of OCD, the

Appendix A: Full Resource Citations and Summaries

social and economic impact of OCD, and the need for intervention studies.

Article ID:	**79**
Citation:	Elsner P, Wigger-Alberti W [2003]. Skin-conditioning products in occupational dermatology. Int Arch Occup Environ Health *76*(5):351–54.
Resource type:	Journal article—review, meta-analysis
Educational materials:	No
Number of references:	20
Industries/occupations:	
Specific process:	
Chemical:	hand cleansers
Specific chemicals:	
Mixtures:	No
Audience:	Professional

Topics addressed:		
	A	Overview
	A.4	Skin physiology and function as barriers to chemical insults
	D	Hazard identification
	D.1	Potential health effects resulting from specific chemicals
	D.1.A	Irritant contact dermatitis
	F	Risk management
	F.1	Exposure control strategies
	F.1.D	PPE and PPE rules
	F.1.E	Skin management, barrier creams, moisturizers, cleansers, and rubs

Summary:	This review discusses the chemistry and mode of action of moisturizers in their prevention of occupational contact dermatitis.

Article ID:	**80**
Citation:	Emmett EA [2002]. Occupational contact dermatitis I: Incidence and return to work pressures. Am J Contact Dermat *13*(1).
Resource type:	Journal article—review, meta-analysis
Educational materials:	No

Appendix A: Full Resource Citations and Summaries

Number of references:	8
Industries/occupations:	
Specific process:	
Chemical:	
Specific chemicals:	
Mixtures:	No
Audience:	Professional
Topics addressed:	B Surveillance and clinical aspects
	B.1 Surveillance study reporting incidences of occupational skin exposures
	B.1.A Skin exposure major focus
	B.2 Loss of workdays and impact on productivity
Summary:	This is the first article in a two-part series. Both articles are summarized below:

Part 1 (ID 80): Describes changes in the incidence of recorded occupational skin disease from 1972 to 1999 and suggests explanations for periodic changes. It provides trends for some specific industries. The authors argue for a more sophisticated approach to prevention and management to reduce the burden of occupational skin disease.

Part 2 (ID 81): Presents the present state of risk assessment (including components of hazard identification, dermal exposure measurement, absorption, dose-response, and risk characterization) and the prognosis for occupational contact dermatitis.

Article ID:	**81**
Citation:	Emmett EA [2003]. Occupational contact dermatitis II: Risk assessment and prognosis. Am J Contact Dermat *14*(1).
Resource type:	Journal article—review, meta-analysis
Educational materials:	No
Number of references:	51
Industries/occupations:	General—overview
Specific process:	
Chemical:	General—overview
Specific chemicals:	
Mixtures:	No
Audience:	Professional

Indexed Dermal Bibliography

Appendix A: Full Resource Citations and Summaries

Topics addressed:	A	Overview
	A.3	Investigation, intervention, and control of occupational skin exposures
	A.4	Skin physiology and function as barriers to chemical insults
	B	Surveillance and clinical aspects
	B.1	Surveillance study reporting incidences of occupational skin exposures
	B.1.A	Skin exposure major focus
	B.4	Clinical protocols for recognition of skin exposure health effects
	C	Exposure characterization
	C.2	Description of factors influencing exposure conditions
	C.2.B	Exposure concentration
	C.2.C	Skin area affected
	C.2.E	Uptake
	C.4	Direct methods to measure exposure
	C.4.B	Skin
	C.4.C	Biomonitoring
	D	Hazard identification
	D.1	Potential health effects resulting from specific chemicals
	D.1.A	Irritant contact dermatitis
	D.1.B	Allergic contact dermatitis/sensitization
	D.2	Summaries of health effects, dose-response relationships
	E	Risk assessment
	E.1	Guidelines for risk assessment or analysis
21-30	E.1.A	Localized health effects

Summary: This is the second article in a 2-part series. Both articles are summarized below:

Part 1 (ID 80): Describes changes in the incidence of recorded occupational skin disease from 1972 to 1999 and suggests explanations for periodic changes. It also examines trends for some specific industries. The authors argue for a more sophisticated approach to prevention and management to reduce the burden of occupational skin disease.

Part 2 (ID 81): Presents the present state of risk assessment (including components of hazard identification, dermal

Appendix A: Full Resource Citations and Summaries

exposure measurement, absorption, dose-response, and risk characterization) and the prognosis for occupational contact dermatitis.

Article ID:	**82**
Citation:	Enviroderm Services [2005]. Dermatological engineering. [www.enviroderm.co.uk/].
Resource type:	Web site
Educational materials:	Yes
Number of references:	
Industries/occupations:	General—overview
Specific process:	
Chemical:	General—overview, coolants, latex
Specific chemicals:	Metalworking fluids
Mixtures:	No
Audience:	Professional

Topics addressed:

A	Overview
A.1	Occurrence of skin exposures in the workplace
A.2	Health hazards resulting from skin exposure to chemicals
A.4	Skin physiology and function as barriers to chemical insults
B	Surveillance and clinical aspects
B.3	Surveillance study protocols/procedures for gathering data
B.4	Clinical protocols for recognition of skin exposure health effects
C	Exposure characterization
C.1	Workplace factors associated with harmful skin exposures
C.2	Description of factors influencing exposure conditions
C.2.A	Exposure intensity/frequency/duration
C.2.B	Exposure concentration
C.2.C	Skin area affected
C.2.E	Uptake
C.4	Direct methods to measure exposure
C.4.B	Skin
D	Hazard identification

Indexed Dermal Bibliography

Appendix A: Full Resource Citations and Summaries

D.1	Potential health effects resulting from specific chemicals
D.1.A	Irritant contact dermatitis
D.1.B	Allergic contact dermatitis/sensitization
D.1.C	Systemic toxicity
E	Risk assessment
E.1	Guidelines for risk assessment or analysis
E.1.A	Localized health effects
E.1.B	Systemic health effects
E.2	Example of risk assessments
F	Risk management
F.1	Exposure control strategies
F.1.A	Substitution
F.1.B	Engineering controls
F.1.C	Work practice/Administrative controls
F.1.D	PPE and PPE rules
F.1.E	Skin management, barrier creams, moisturizers, cleansers, and rubs
F.2	Protocols for risk management
F.2.B	Development of approach to achieve exposure reduction goal

Summary: Enviroderm Services is a U.K.-based consulting firm founded by Chris Packham, a recognized expert in the field of dermal exposure, that specializes in the workplace dermal exposures prevention and control. Although consulting services are limited to the U.K., they have available through their Web site a different dermal exposure-related materials, including literature, educational and training materials, workplace posters, skin-monitoring equipment, dermal risk assessment forms, and skin health surveillance tools including forms and questionnaires. The Web site also contains a brief description of all the materials and literature available for purchase.

Literature available through the Web site includes:

- *Essentials of Occupational Skin Management.* A 15-chapter textbook featured as a separate resource in this indexed dermal bibliography (see ID 169).
- Risk Assessment (for dermal exposure) Forms
- Technical Bulletin No.1 Skin Management
- Technical Bulletin No.2 Occupational Skin Diseases

Appendix A: Full Resource Citations and Summaries

- Technical Bulletin No.3 Chemical protection using gloves
- Technical Bulletin No.4 Health Surveillance and the skin
- Technical Bulletin No.5 Irritant Contact Dermatitis
- Technical Bulletin No.6 Thoughts on Latex Allergy
- Technical Bulletin No.7 Personal Hygiene
- Technical Bulletin No.8 Is it occupational?
- Technical Bulletin No.9 Emollients
- Technical Bulletin No.10 Barrier Creams
- Technical Bulletin No.11 Risk Assessment for Dermal Exposure
- Technical Bulletin No.12 Risk management
- Technical Bulletin No.13 Allergic Skin Disorders
- Technical Bulletin No.14 Metalworking fluids
- Technical Bulletin No.15 Infection control and the skin
- Technical Bulletin No.16 Investigating a skin problem
- Technical Bulletin No.17 How hazardous is that chemical?
- Technical Bulletin No.18 Skin Exposure Measurement

Article ID:	83
Citation:	Organisation for Economic Co-operation and Development (OECD) [1999]. OECD series on testing and assessment, number 16: detailed review document on classification systems for skin irritation/corrosion in OECD member countries. Paris, France: OECD, ENV/JM/MONO(99)6.
Resource type:	Journal article—review, meta-analysis
Educational materials:	No
Number of references:	22
Industries/occupations:	
Specific process:	
Chemical:	
Specific chemicals:	
Mixtures:	No
Audience:	Professional
Topics addressed:	A Overview
	A.5 Dermal regulations and skin notations

Indexed Dermal Bibliography

Appendix A: Full Resource Citations and Summaries

Summary:	This report compares the dermal irritation/corrosion hazard classification procedures used in Canada, the US, OECD, European Union, and Norway. Issues requiring resolution are discussed.

Article ID:	84	
Citation:	OECD [2004]. OECD series on testing and assessment, number 28: Guidance document for the conduct of skin absorption studies. Paris, France: OECD, ENV/JM/MONO (2004)2.	
Resource type:	Guideline	
Educational materials:	No	
Number of references:	59	
Industries/occupations:		
Specific process:		
Chemical:		
Specific chemicals:		
Mixtures:	No	
Audience:	Professional	
Topics addressed:	C	Exposure characterization
	C.2	Description of factors influencing exposure conditions
	C.2.E	Uptake
	D	Hazard identification
	D.3	Characterization protocols
	D.3.E	Measurement of skin permeation rates and reservoir effects
Summary:	The OECD *Guidance Document for the Conduct of Skin Absorption Studies* was published by the Organization for Economic Co-operation and Development (OECD), an intergovernmental organization which representatives 30 industrialized countries in North America, Europe, and the Pacific, as well as the European Commission. It provides guidance developed by OECD at Research Triangle Park, North Carolina, in October 1997 called "Percutaneous Absorption Methods as Test Guidelines" for *in vitro* and *in vivo* studies.	

Article ID:	85
Citation:	USEPA [1998]. Harmonized test guidelines. [www.epa.gov/opptsfrs/home/guidelin.htm].

Appendix A: Full Resource Citations and Summaries

Resource type:	Web page
Educational materials:	No
Number of references:	
Industries/occupations:	General—overview, Agricultural
Specific process:	
Chemical:	General—overview, pesticides
Specific chemicals:	
Mixtures:	No
Audience:	Professional
Topics addressed:	C Exposure characterization
	C.4 Direct methods to measure exposure
	C.4.A Surfaces
	C.4.B Skin
	D Hazard identification
	D.3 Characterization protocols
	D.3.B Irritation potential
	D.3.C Sensitization potential
	D.3.D Potential to cause systemic effects
	D.3.E Measurement of skin permeation rates and reservoir effects

Summary: The USEPA Office of Prevention, Pesticides and Toxic Substances (OPPTS) harmonized test guidelines were developed to minimize variations in testing procedures under the Toxic Substances Control Act (TSCA) and the FIFRA. These were developed primarily for occupational pesticides and other toxic substances.

In Series 870, *The Health Effects Test Guidelines*, guidelines can be found for use in the testing of pesticides and toxic substances and the development of test data that must be submitted to the USEPA for review under federal regulations. The following dermal-related test guidelines can be found in this series:

- 870.1200 Acute dermal toxicity
- 870.2500 Acute dermal irritation
- 870.2600 Skin sensitization
- 870.3200 21/28-Day dermal toxicity
- 870.3250 90-Day dermal toxicity

In Series 875, *The Occupational and Residential Exposure Test Guidelines: Post Application Exposure Guidelines*, the following dermal exposure test guidelines can be found:

Appendix A: Full Resource Citations and Summaries

- 875.1100 Dermal exposure-outdoor
- 875.1200 Dermal exposure-indoor
- 875.2400 Dermal exposure

Article ID:	86
Citation:	USEPA [2000]. Summary report for the workshop on issues associated with dermal exposure and uptake. Washington, DC: U.S. Environmental Protection Agency, FRL-7032-8.
Resource type:	Technical publication/report
Educational materials:	No
Number of references:	22
Industries/occupations:	Waste Management
Specific process:	
Chemical:	
Specific chemicals:	
Mixtures:	No
Audience:	Professional
Topics addressed:	C Exposure characterization
	C.2 Description of factors influencing exposure conditions
	C.2.E Uptake
	C.5 Exposure modeling
Summary:	This paper is a summary of a December 1998 workshop that discussed issues concerning dermal uptake, permeability, and absorbed dose to chemicals.

Article ID:	87
Citation:	USEPA [2005]. [www.epa.gov].
Resource type:	Web site
Educational materials:	No
Number of references:	
Industries/occupations:	General—overview, Agricultural, Manufacturing
Specific process:	autobody painting
Chemical:	General—overview, pesticides, petroleum products & lubricants, solvents
Specific chemicals:	

Appendix A: Full Resource Citations and Summaries

Mixtures:	No	
Audience:	Professional	
Topics addressed:	B	Surveillance and clinical aspects
	B.1	Surveillance study reporting incidences of occupational skin exposures
	B.1.A	Skin exposure major focus
	B.2	Loss of workdays and impact on productivity
	B.3	Surveillance study protocols/procedures for gathering data
	C	Exposure characterization
	C.2	Description of factors influencing exposure conditions
	C.2.A	Exposure intensity/frequency/duration
	C.2.E	Uptake
	C.4	Direct methods to measure exposure
	C.4.A	Surfaces
	C.4.B	Skin
	C.4.C	Biomonitoring
	C.5	Exposure modeling
	D	Hazard identification
	D.3	Characterization protocols
	D.3.B	Irritation potential
	D.3.C	Sensitization potential
	D.3.D	Potential to cause systemic effects
	D.3.E	Measurement of skin permeation rates and reservoir effects
	E	Risk assessment
	E.1	Guidelines for risk assessment or analysis
	E.1.A	Localized health effects
	E.1.B	Systemic health effects
	E.2	Example of risk assessments
	F	Risk management
	F.1	Exposure control strategies
	F.1.A	Substitution
	F.1.C	Work practice/Administrative controls
	F.1.D	PPE and PPE rules
Summary:	The USEPA is the federal agency tasked with protecting human health and the environment. In addition to developing and enforcing regulations, the USEPA also	

Indexed Dermal Bibliography

Appendix A: Full Resource Citations and Summaries

performs environmental research, sponsors voluntary partnerships and programs, advances environmental education, and publishes information associated with the environment. Although the USEPA's work deals more with environmental exposures than occupational exposures (except in the case of pesticides), some resources are designed for occupational settings and some are applicable to both. The USEPA has produced resources related to dermal exposure to chemicals that can be found on their Web site, including:

- Series 875, *Occupational and Residential Exposure Test Guidelines: Post Application Exposure Guidelines, Group B*. These guidelines provide background information on the application of exposure monitoring test guidelines, dermal exposure (outdoor), dermal exposure (indoor), biological monitoring, and data reporting and calculations.

- *Cleaner Technologies Substitutes Assessment: A Methodology and Resource Guide*. This contains guidelines on evaluating chemical substitution, though not specific to dermal exposures. Recommendations that may be applicable can be found here.

- "Choosing the Right Gloves for Painting Cars." This covers how to select chemical resistant gloves for automobile paint work.

- *Dermal Exposure Assessment: Principles and Applications*. This 1992 guidance document covers the principles of dermal absorption and outlines procedures on how to apply these principles to actual dermal exposure assessments involving contact with chemical vapors, air, soil and water.

- *Summary Report for the Workshop on Issues Associated With Dermal Exposure and Uptake*. This is a summary from a 1998 workshop that discussed technical issues associated with dermal exposure and risk assessment.

- *Exposure Factors Handbook*, Chapter 6, "Dermal." This 1997 document on general dermal exposure considerations is directed at environmental exposures, but may apply to some occupational settings as well.

Article ID:	**88**
Citation:	European Agency for Safety and Health and Work [2005]. European Agency for Safety and Health and Work. [http://europe.osha.eu.int/OSHA].

Appendix A: Full Resource Citations and Summaries

Resource type:	Web site
Educational materials:	No
Number of references:	
Industries/occupations:	General—overview
Specific process:	
Chemical:	General—overview
Specific chemicals:	
Mixtures:	No
Audience:	General

Topics addressed:

A	Overview
A.1	Occurrence of skin exposures in the workplace
A.2	Health hazards resulting from skin exposure to chemicals
A.3	Dermal regulations and skin notations
B	Exposure characterization
B.1	Job/tasks, industries/processes, or chemicals associated with skin exposures
C	Hazard identification
C.2	Tables/charts/lists of hazards for specific chemicals
E	Risk management
E.3	"Best practices"/guidelines/recommendations
E.3.A	Substitution
E.3.B	Engineering controls
E.3.C	Work practice/administration controls
E.3.D	PPE and PPE rules
E.3.E	Skin management, barrier creams, moisturizers, cleansers, and rubs

Summary: The European Agency for Safety and Health at Work collects, analyzes, and promotes occupational safety- and health-related information in Europe. The agency is a tripartite European Union organization that brings together representatives from governments, employers' and workers' organizations, as well as from the European Commission. The agency's web portal provides links to over 30 national Web sites, usually the lead OSH organization in the European Union. member states, candidate countries, and other international partners. Information on dermal exposure can be found here based on chemical, risks, industry or sector, and topics of interest. Information is also available in multiple languages.

Appendix A: Full Resource Citations and Summaries

Useful information on dermal exposures available through the agency's Web site includes the following fact sheets:

- Issue 34, *Eliminating and Substituting Dangerous Substances*
- Issue 35, *Communicating Information about Dangerous Substance*
- Issue 40, *Skin Sensitizers*

Article ID:	**89**
Citation:	European Center for Ecotoxicity and Toxicology of Chemicals (ECETOC) [1999]. Skin and respiratory sensitisers: Reference chemical data bank. Brussels, Belgium: ECETOC, Technical Report No. 77.
Resource type:	Technical publication/report
Educational materials:	No
Number of references:	292
Industries/occupations:	
Specific process:	
Chemical:	
Specific chemicals:	
Mixtures:	No
Audience:	Professional
Topics addressed:	C Exposure characterization
	C.5 Exposure modeling
	D Hazard identification
	D.1 Potential health effects resulting from specific chemicals
	D.1.A Irritant contact dermatitis
	D.1.B Allergic contact dermatitis/sensitization
Summary:	This paper provides a list of skin and respiratory sensitizers which may be used for the validation of *in vivo* or *in vitro* toxicological tests. The list also identifies chemicals that will facilitate the evaluation and validation of proposed predictive test methods for skin and/or respiratory sensitization potential. It documents those chemicals that are recommended for use as positive and negative controls in the assessment of new predictive tests for skin or respiratory sensitization potential and assesses the utility and accuracy of "novel" test methods.

Appendix A: Full Resource Citations and Summaries

Article ID:	90
Citation:	Fehrenbacher C, Arnold F, Marquart H [2003]. Approaches for occupational dermal exposure assessment and management. In: DiNardi SR, ed. The occupational environment: its evaluation, control, and management. 2nd ed. Fairfax, VA: AIHA.
Resource type:	Book/monograph, chapter
Educational materials:	No
Number of references:	53
Industries/occupations:	
Specific process:	
Chemical:	
Specific chemicals:	2,4-dichlorophenol
Mixtures:	No
Audience:	Professional

Topics addressed:

C	Exposure characterization
C.2	Description of factors influencing exposure conditions
C.2.A	Exposure intensity/frequency/duration
C.2.B	Exposure concentration
C.2.C	Skin area affected
C.2.E	Uptake
C.4	Direct methods to measure exposure
C.4.A	Surfaces
C.4.B	Skin
C.4.C	Biomonitoring
C.5	Exposure modeling
E	Risk assessment
E.1	Guidelines for risk assessment or analysis
E.1.A	Localized health effects
E.1.B	Systemic health effects
F	Risk management
F.1	Exposure control strategies
F.1.A	Substitution
F.1.B	Engineering controls
F.1.C	Work practice/Administrative controls
F.1.D	PPE and PPE rules
F.1.E	Skin management, barrier creams, moisturizers, cleansers, and rubs

Appendix A: Full Resource Citations and Summaries

Summary:	Chapter 17, "Approaches for Occupational Dermal Exposure Assessment and Management," from the AIHA book *The Occupational Environmental: Its Evaluation Control and Management (the White Book)*. This chapter discusses dermal exposure monitoring methods, the process of dermal absorption, methods to measure dermal uptake, a tiered approach to performing dermal exposure assessments, and the control and management of occupational dermal exposures.

Article ID:	91	
Citation:	Fenske RA [1993]. Dermal exposure assessment techniques. Ann Occup Hyg 37(6):687–706.	
Resource type:	Journal article—review, meta-analysis	
Educational materials:	No	
Number of references:	94	
Industries/occupations:		
Specific process:		
Chemical:		
Specific chemicals:		
Mixtures:	No	
Audience:	Professional	
Topics addressed:	C	Exposure characterization
	C.2	Description of factors influencing exposure conditions
	C.2.A	Exposure intensity/frequency/duration
	C.2.B	Exposure concentration
	C.2.C	Skin area affected
	C.2.E	Uptake
	C.4	Direct methods to measure exposure
	C.4.A	Surfaces
	C.4.B	Skin
	C.4.C	Biomonitoring
	C.5	Exposure modeling
Summary:	This article discusses three primary pathways to exposure: immersion, deposition of aerosol or uptake of vapor through the skin, and contact with contaminated surfaces. It also discusses three primary sampling methods: surrogate skin, chemical removal, and fluorescent tracers. This article also presents a dermal exposure sampling strategy which addresses issues associated with the	

sampling method, representative sampling, and sample duration. Finally, it recommends the development of dermal occupational exposure limits (DOELs) for selected workplaces and chemical agents.

Article ID:	92
Citation:	Fenske RA [2000]. Dermal exposure: A decade of real progress. Ann Occup Hyg *44*(7):489–91.
Resource type:	Other—editorial
Educational materials:	No
Number of references:	17
Industries/occupations:	
Specific process:	
Chemical:	
Specific chemicals:	
Mixtures:	No
Audience:	Professional
Topics addressed:	A Overview
	A.3 Investigation, intervention, and control of occupational skin exposures
Summary:	This letter to the editor introduces an issue relevant to the journal which features recent work on workplace dermal exposure assessment. It summarizes the history of progress made in dermal exposure assessment through 2000.

Article ID:	93
Citation:	Fitzpatrick D, Corish J, Hayes B [2004]. Modeling skin permeability in risk assessment—the future. Chemosphere *55*(10):1309–14.
Resource type:	Journal article—review, meta-analysis
Educational materials:	No
Number of references:	31
Industries/occupations:	
Specific process:	
Chemical:	
Specific chemicals:	
Mixtures:	No
Audience:	Professional
Topics addressed:	C Exposure characterization

Appendix A: Full Resource Citations and Summaries

	C.2	Description of factors influencing exposure conditions
	C.2.E	Uptake
	C.5	Exposure modeling
	D	Hazard identification
	D.3	Characterization protocols
	D.3.E	Measurement of skin permeation rates and reservoir effects
	D.3.F	QSARs—development, validation, and application
Summary:	This article presents recent progress in skin permeability modeling and compares two methods of assessing skin permeability: quantitative structure-activity relationships (QSARs) and mathematical modeling based on analytical or numerical solutions to partition and transport equations. It also proposes steps that can be taken for future advancements in this field.	

Article ID:	94	
Citation:	Flynn MR, Koto Y, Fent K, Nylander-French LA [2006]. Modeling dermal exposure—an illustration for spray painting applications. J Occup Environ Hyg 3(9):475–80.	
Resource type:	Journal article—primary	
Educational materials:	No	
Number of references:	15	
Industries/occupations:	Construction, Automotive	
Specific process:	Spray painting, Autobody workers	
Chemical:		
Specific chemicals:		
Mixtures:	No	
Audience:	Professional	
Topics addressed:	C	Exposure characterization
	C.5	Exposure modeling
Summary:	This study presents a model to ascertain dermal exposure from aerosol spray paint deposition on human forearm hair.	

Article ID:	95
Citation:	Forsberg K, Mansdorf SZ [2002]. Quick selection guide to CPC, 4th ed. New York: J. Wiley.

Appendix A: Full Resource Citations and Summaries

Resource type:	Brochure, pamphlet
Educational materials:	Yes
Number of references:	0
Industries/occupations:	
Specific process:	
Chemical:	
Specific chemicals:	Index includes over 700 chemicals
Mixtures:	No
Audience:	General

Topics addressed:

A	Overview
A.3	Dermal regulations and skin notations
C	Hazard identification
C.1	Risk phrases, hazard symbols, skin designations
C.2	Tables/charts/lists of hazards for specific chemicals
E	Risk management
E.3	"Best practices"/guidelines/recommendations
E.3.D	PPE and PPE rules

Summary: This pocket-sized field guide for spill responders, safety engineers, industrial hygienists, chemists and chemical engineers, and other workers presents information on 700 chemicals, additional synonyms, CAS numbers, risk codes, and special notations to alert the user. It also discusses 16 PPE barrier materials used.

CONTENTS

1. Introduction
2. Selection and Use of CPC.
3. Chemical Index. Contains the chemical class numbers, chemical names and synonyms, chemical abstract service (CAS) numbers, risk codes, and special "skin" and "caution" notations.
4. Selection Recommendations. Provides color coded recommendations organized numerically by chemical class based on 11,000 permeation and 3,000 degradation test data.
5. Glossary
6. Standards for CPC
7. Manufacturers of CPC. Includes the names, addresses, and phone numbers of the suppliers and manufacturers of the CPC materials.

Indexed Dermal Bibliography

Appendix A: Full Resource Citations and Summaries

Article ID:	96
Citation:	Furtaw EJ Jr [2001]. An overview of human exposure modeling activities at the USEPA's National Exposure Research Laboratory (NERL). Toxicol Ind Health *17*:302–14.
Resource type:	Journal article—primary
Educational materials:	No
Number of references:	18
Industries/occupations:	
Specific process:	
Chemical:	
Specific chemicals:	
Mixtures:	No
Audience:	Professional
Topics addressed:	C Exposure characterization
	C.5 Exposure modeling
	E Risk assessment
	E.2 Example of risk assessments

Summary: This article reviews the following risk assessment models developed by the USEPA's NERL. NERL modeling efforts, though directed at environmental exposures, are applicable to occupational exposures as well. Modeling has focused on understanding the factors that influence exposure and has been designed for use in risk assessments and for risk management. Specific models reviewed include:

- Community Multiscale Air Quality (CMAQ) Model (Models-3/CMAQ) for pollutant concentrations in ambient (outdoor) air.
- Computational fluid dynamics (CFD) for air flow and pollutant concentrations.
- Stochastic Human Exposure and Dose Simulation (SHEDS) Model for human inhalation exposure to airborne particulates, toxics, or pesticides.
- Framework for Risk Analysis in Multimedia Environmental Systems—Multimedia, Multipathway, Multireceptor Risk Assessment (FRAMES-3MRA) for human and ecological exposure and risk assessments of hazardous waste sites.
- Exposure-Related Dose-Estimating Model (ERDEM) for physiologically based pharmacokinetic (PBPK) modeling of pesticides and VOCs.

Appendix A: Full Resource Citations and Summaries

Article ID:	**97**
Citation:	Garrod AN, Rajan-Sithamparanadarajah R [2003]. Developing COSHH essentials: Dermal exposure, personal protective equipment and first aid. Ann Occup Hyg 47(7):577–88.
Resource type:	Journal article—review, meta-analysis
Educational materials:	No
Number of references:	20
Industries/occupations:	
Specific process:	
Chemical:	
Specific chemicals:	
Mixtures:	No
Audience:	Professional
Topics addressed:	A Overview
	A.5 Dermal regulations and skin notations
	C Exposure characterization
	C.2 Description of factors influencing exposure conditions
	C.2.A Exposure intensity/frequency/duration
	C.2.B Exposure concentration
	C.2.E Uptake
	C.5 Exposure modeling
Summary:	This paper discusses how to apply COSHH Essentials, originally developed in the U.K. to control inhalation exposures in the workplace and to control dermal exposures. It examines the factors affecting skin exposure, and outlines options to band chemical hazards for emergency planning according to a minimum of information, i.e., the danger symbol on a product label. It also discusses dermal hazard classifications.

Article ID:	**98**
Citation:	Geer LA, Anna D, Curbow B, ener-West M, de Joode BW, Mitchell C, Buckley TJ [2007]. Survey assessment of worker dermal exposure and underlying behavioral determinants. J Occup Environ Hyg 4(11):809–20.
Resource type:	Journal article—primary
Educational materials:	No
Number of references:	34

Appendix A: Full Resource Citations and Summaries

Industries/occupations:	
Specific process:	
Chemical:	
Specific chemicals:	
Mixtures:	No
Audience:	Professional
Topics addressed:	F Risk management
	F.1 Exposure control strategies
	F.1.C Work practice/Administrative controls
	F.1.D PPE and PPE rules
Summary:	This study assesses worker knowledge, attitudes, and perceptions of workplace dermal hazards using a questionnaire, compares worker and manager scores, evaluates worker dermal exposure using DREAM, and identifies potential behavioral factors underlying exposure.

Article ID:	**99**
Citation:	Gerberick GF, Ryan CA, Kern PS, Dearman RJ, Kimber I, Patlewicz GY, Basketter DA [2004]. A chemical dataset for evaluation of alternative approaches to skin-sensitization testing. Contact Dermatitis 50(5):274–88.
Resource type:	Journal article—review, meta-analysis
Educational materials:	No
Number of references:	45
Industries/occupations:	
Specific process:	
Chemical:	
Specific chemicals:	
Mixtures:	No
Audience:	Professional
Topics addressed:	D Hazard identification
	D.1 Potential health effects resulting from specific chemicals
	D.1.B Allergic contact dermatitis/sensitization
	D.3 Characterization protocols
	D.3.C Sensitization potential
Summary:	This article presents a list of 244 chemicals and their relative sensitization potency, as determined by the local lymph node assay (LLNA). The authors state this dataset

Appendix A: Full Resource Citations and Summaries

can be used to evaluate and calibrate novel approaches to skin sensitization testing.

Article ID:	**100**
Citation:	Goede HA, Tijssen SC, Schipper HJ, Warren N, Oppl R, Kalberlah F, van Hemmen JJ [2003]. Classification of dermal exposure modifiers and assignment of values for a risk assessment toolkit. Ann Occup Hyg 47(8):609–18.
Resource type:	Journal article—review, meta-analysis
Educational materials:	No
Number of references:	39
Industries/occupations:	
Specific process:	
Chemical:	
Specific chemicals:	
Mixtures:	No
Audience:	Professional
Topics addressed:	E Risk assessment
	E.1 Guidelines for risk assessment or analysis
	E.1.A Localized health effects
	E.1.B Systemic health effects
Summary:	This article is the 4th article of a six-part series on RISKOFDERM, a tool for conducting risk assessments. The series was published in the *Annals of Occupational Hygiene* in 2003. The following briefly summarizes each paper in the series:

1. ID 212—Outlines a "toolkit" for conducting dermal occupational risk assessments.
2. ID 163—Describes the assumptions in the toolkit and describes an approach to exposure assessment used by the toolkit.
3. ID 139—Describes the determinants relevant for dermal exposure models in the scope of regulatory risk assessment.
4. ID 219—Describes how default dermal exposure values can be adjusted for specific work situations.
5. ID 100—Describes the derivation of the toolkit's default task-based dermal exposure values.
6. ID 193—Describes the development of "intrinsic toxicity" (IT) scores used for hazard characterization.

Appendix A: Full Resource Citations and Summaries

Article ID:	**101**
Citation:	Graves CG, Matanoski GM, Tardiff RG [2000]. Carbonless copy paper and workplace safety: a review. Regul Toxicol Pharmacol 32(1):99–117.
Resource type:	Journal article—review, meta-analysis
Educational materials:	No
Number of references:	122
Industries/occupations:	
Specific process:	
Chemical:	Cleaning agents, PCBs, other: carbonless copy paper (CCP)
Specific chemicals:	Formaldehyde
Mixtures:	No
Audience:	Professional
Topics addressed:	A Overview
	A.1 Occurrence of skin exposures in the workplace
	A.2 Health hazards resulting from skin exposure to chemicals
	B Surveillance and clinical aspects
	B.1 Surveillance study reporting incidences of occupational skin exposures
	B.1.A Skin exposure major focus
	D Hazard identification
	D.1 Potential health effects resulting from specific chemicals
	D.1.A Irritant contact dermatitis
	D.3 Characterization protocols
	D.3.B Irritation potential
	D.3.C Sensitization potential
Summary:	This paper presents a meta-analysis of 121 papers published on CCP since 1987. CCP has been alleged to cause skin irritation, however, this weight-of-evidence analysis concludes that no irritation or sensitization from CCP should be expected under normal conditions of manufacture and use.

Article ID:	**102**
Citation:	University of California, San Diego [1998]. Prediction and assessment of dermal exposure. La Jolla, CA: University of California, San Diego, AD-a358 903.1998.

Appendix A: Full Resource Citations and Summaries

Resource type:	Technical publication/report
Educational materials:	No
Number of references:	13
Industries/occupations:	
Specific process:	
Chemical:	
Specific chemicals:	
Mixtures:	No
Audience:	Professional
Topics addressed:	C Exposure characterization
	C.2 Description of factors influencing exposure conditions
	C.2.E Uptake
	C.5 Exposure modeling
Summary:	This paper presents the basis for algorithms developed to predict the rate of absorption of chemicals following dermal exposure. These algorithms are necessary for performing risk assessments. The paper includes the results of a literature review as well as the algorithm test results.

Article ID:	103
Citation:	Hamann CP, DePaola LG, Rodgers PA [2005]. Occupation-related allergies in dentistry. J Am Dent Assoc *136*(4):500–10.
Resource type:	Journal article—review, meta-analysis
Educational materials:	No
Number of references:	73
Industries/occupations:	Dentistry
Specific process:	
Chemical:	Latex, plastics and resins, rubber additives, other: adhesives, antiseptics, artificial fingernails, dental bonding agents, disinfectants, equipment sterilization solutions, skin care products, rubber gloves, radiographic and photo chemical
Specific chemicals:	
Mixtures:	No
Audience:	Professional
Topics addressed:	A Overview
	A.1 Occurrence of skin exposures in the workplace

Indexed Dermal Bibliography

Appendix A: Full Resource Citations and Summaries

A.2	Health hazards resulting from skin exposure to chemicals
B	Surveillance and clinical aspects
B.1	Surveillance study reporting incidences of occupational skin exposures
B.1.A	Skin exposure major focus
D	Hazard identification
D.1	Potential health effects resulting from specific chemicals
D.1.A	Irritant contact dermatitis
D.1.B	Allergic contact dermatitis/sensitization
F	Risk management
F.1	Exposure control strategies
F.1.A	Substitution
F.1.B	Engineering controls
F.1.C	Work practice/Administrative controls
F.1.D	PPE and PPE rules
F.1.E	Skin management, barrier creams, moisturizers, cleansers, and rubs

Summary: This paper presents the health effects associated with occupation-related allergies in dentistry. Natural rubber latex (NRL) protein allergy, allergic contact dermatitis, and irritant contact dermatitis are discussed. Topics include diagnosis, exposure measuring, management and prevention, and some surveillance information.

Article ID:	**104**
Citation:	Hatch KL, Maibach HI [2000]. Textile dye allergic contact dermatitis prevalence. Contact Dermatitis 42(4):187–95.
Resource type:	Journal article—review, meta-analysis
Educational materials:	No
Number of references:	20
Industries/occupations:	Manufacturing
Specific process:	
Chemical:	Organic dyes
Specific chemicals:	Paper examines over 60 dyes for prevalence data including disperse orange 3, yellow 3, red 1, blue 124, black 1, brown 1, and reactive green 12.
Mixtures:	No
Audience:	Professional

Appendix A: Full Resource Citations and Summaries

Topics addressed:	B	Surveillance and clinical aspects
	B.1	Surveillance study reporting incidences of occupational skin exposures
	B.1.A	Skin exposure major focus
	C	Exposure characterization
	C.4	Direct methods to measure exposure
	C.4.B	Skin
Summary:		This paper summarizes textile dye allergic contact dermatitis prevalence studies and makes recommendations for future work.

Article ID:	**105**
Citation:	HSE [2005]. Skin at work. [www.hse.gov.uk/skin/].
Resource type:	Web page
Educational materials:	No
Number of references:	
Industries/occupations:	General—overview, Beauty/Cosmetology, Manufacturing—Chemical, Manufacturing—Other, Service—Food
Specific process:	Hairdressers, catering, printing
Chemical:	General—overview, food products, heavy metals/inorganic compounds, latex, nanoparticles, pesticides, plastics and resins, PCBs, soaps and detergents, solvents
Specific chemicals:	Isocyanates, epoxy resins
Mixtures:	No
Audience:	Professional

Topics addressed:	A	Overview
	A.1	Occurrence of skin exposures in the workplace
	A.2	Health hazards resulting from skin exposure to chemicals
	A.3	Investigation, intervention, and control of occupational skin exposures
	A.4	Skin physiology and function as barriers to chemical insults
	B	Surveillance and clinical aspects
	B.1	Surveillance study reporting incidences of occupational skin exposures
	B.1.A	Skin exposure major focus
	B.2	Loss of workdays and impact on productivity

Indexed Dermal Bibliography

Appendix A: Full Resource Citations and Summaries

B.4	Clinical protocols for recognition of skin exposure health effects
C	Exposure characterization
C.1	Workplace factors associated with harmful skin exposures
C.2	Description of factors influencing exposure conditions
C.2.C	Skin area affected
C.4	Direct methods to measure exposure
C.4.A	Surfaces
C.4.B	Skin
C.4.C	Biomonitoring
D	Hazard identification
D.1	Potential health effects resulting from specific chemicals
D.1.A	Irritant contact dermatitis
D.1.B	Allergic contact dermatitis/sensitization
D.2	Summaries of health effects, dose-response relationships
D.3	Characterization protocols
D.3.A	Corrosivity
D.3.C	Sensitization potential
D.3.E	Measurement of skin permeation rates and reservoir effects
E	Risk assessment
E.1	Guidelines for risk assessment or analysis
E.1.A	Localized health effects
E.1.B	Systemic health effects
E.2	Example of risk assessments
F	Risk management
F.1	Exposure control strategies
F.1.B	Engineering controls
F.1.C	Work practice/Administrative controls
F.1.D	PPE and PPE rules
F.1.E	Skin management, barrier creams, moisturizers, cleansers, and rubs

Summary: Britain's Health and Safety Commission (HSC) and HSE are responsible for the regulation of occupational health and safety risks arising in the U.K. Any regulatory information provided here is specific to the U.K. This Web

Appendix A: Full Resource Citations and Summaries

site provides comprehensive health and safety information, with most dermal exposure-related information found on the topic page *Skin at Work*.

The HSE's *Skin at Work* Web page includes a variety of free leaflets, including: *Preventing Dermatitis at Work, Advice for Employers and Employees;* MS24—*Medical Aspects of Occupational Skin Disease;* Managing Health and Safety topics, *Personal Protective Equipment (PPE)*, *Risk Assessment,* and a number of chemical-specific leaflets. It also contains links to chemical-specific alert notices. The *Skin at Work* Web page provides specific dermal exposure information for the following industries: hairdressing, catering, and printing.

The HSE Web site also has information on different dermal exposure research topics, including:

- Contact dermatitis
- Occupational dermatitis
- Skin disease surveillance data
- Development of a method to assess biologically relevant dermal exposure
- Dermal exposure resulting from liquid contamination
- *In vitro* dermal absorption of liquids
- Health effects of particles produced for nanotechnologies
- Pesticides in air and/or on surfaces
- Draft guidelines on route-to-route extrapolation of toxicity data when assessing health risks of chemicals
- Development of a field method for the assessment of the effectiveness of barrier creams in preventing skin irritation reactions
- An assessment of skin sensitization by the use of epoxy resin in the construction industry

Article ID:	**106**
Citation:	Hewett P [2001]. Misinterpretation and misuse of exposure limits. Appl Occup Environ Hyg *16*(2):251–56.
Resource type:	Journal article—review, meta-analysis
Educational materials:	No
Number of references:	28
Industries/occupations:	
Specific process:	

Appendix A: Full Resource Citations and Summaries

Chemical:		
Specific chemicals:		
Mixtures:	No	
Audience:	Professional	
Topics addressed:	A	Overview
	A.5	Dermal regulations and skin notations
	E	Risk assessment
	E.1	Guidelines for risk assessment or analysis
	E.1.A	Localized health effects
	E.1.B	Systemic health effects
	F	Risk management
	F.2	Protocols for risk management
	F.2.B	Development of approach to achieve exposure reduction goal
Summary:	This article discusses occupational exposure limits (OELs) and distinguishes between how they should be used in risk assessment and exposure risk management and how they can be misused in each.	

Article ID:	**107**	
Citation:	Hostynek JJ [2003]. Factors determining percutaneous metal absorption. Food Chem Toxicol 41(3):327–45.	
Resource type:	Journal article—review, meta-analysis	
Educational materials:	No	
Number of references:	157	
Industries/occupations:		
Specific process:		
Chemical:	Heavy metals/inorganic compounds	
Specific chemicals:		
Mixtures:	No	
Audience:	Professional	
Topics addressed:	C	Exposure characterization
	C.2	Description of factors influencing exposure conditions
	C.2.E	Uptake
Summary:	This review article describes factors affecting the permeability of metals through the skin including dose, vehicle, volume, counter ion, chemical bond and polarity, valence, protein reactivity, solubility, age of skin,	

Appendix A: Full Resource Citations and Summaries

anatomical site, homeostatic controls, skin layers, and oxidation/reduction.

Article ID:	**108**
Citation:	USEPA [1998]. Dermal and non-dietary ingestion exposure workshop: NERL Human Exposure Research Program. Research Triangle Park, NC: USEPA, NERL, Human Exposure & Atmospheric Sciences Division, Human Exposure Analysis Branch.
Resource type:	Technical publication/report
Educational materials:	No
Number of references:	125
Industries/occupations:	
Specific process:	
Chemical:	pesticides
Specific chemicals:	
Mixtures:	No
Audience:	Professional
Topics addressed:	C Exposure characterization
	C.2 Description of factors influencing exposure conditions
	C.2.A Exposure intensity/frequency/duration
	C.2.B Exposure concentration
	C.2.C Skin area affected
	C.2.E Uptake
	C.4 Direct methods to measure exposure
	C.4.A Surfaces
	C.4.B Skin
	C.4.C Biomonitoring
	C.5 Exposure modeling
Summary:	This paper presents a summary of the dermal and nondietary ingestion exposure workshop sponsored by USEPA NERL on September 17, 1998. The workshop evaluated methods for measuring and assessing children's exposures to pesticides via dermal contact with contaminated surfaces and objects, and through nondietary ingestion. The workshop also evaluated methods for characterizing concentrations of pesticides on surfaces and quantifying the transfer of contaminants to the skin surface or mouth. The workshop's objectives

Appendix A: Full Resource Citations and Summaries

included identification of exposure assessment methods, determination of best approach, and evaluation of these methods' and approach's strengths and weaknesses. Dermal assessment methods reviewed include the microactivity approach, the macroactivity approach, biomonitoring, passive dosimetry, environmental exposure and activity pattern, florescent tracer, and dermal wash/rinse/wipe.

Article ID:	**109**
Citation:	Human Exposure Research Organisations Exchange (HEROX) [2005]. [www.herox.org/].
Resource type:	Web site
Educational materials:	No
Number of references:	
Industries/occupations:	General—overview
Specific process:	
Chemical:	General—overview
Specific chemicals:	
Mixtures:	No
Audience:	Professional
Topics addressed:	E Risk assessment
	E.1 Guidelines for risk assessment or analysis
	E.1.A Localized health effects
	E.1.B Systemic health effects
Summary:	HEROX is a forum for people interested in research on human exposure to hazardous substances. It provides information related to exposure to carcinogens, dermal exposure assessment, development of analytical methods, and exposure modeling research as well as access to databases on workplace exposure. Material on this site is edited by the Department of Environmental and Occupational Medicine at the University of Aberdeen, UK.

Article ID:	**110**
Citation:	Interagency Coordinating Committee on the Validation of Alternative Methods (ICCVAM) [2005]. [http://iccvam.niehs.nih.gov].
Resource type:	Web site
Educational materials:	No
Number of references:	

Appendix A: Full Resource Citations and Summaries

Industries/occupations:		
Specific process:		
Chemical:	General—overview	
Specific chemicals:		
Mixtures:	No	
Audience:	Professional	
Topics addressed:	D	Hazard identification
	D.1	Potential health effects resulting from specific chemicals
	D.1.B	Allergic contact dermatitis/sensitization
	D.3	Characterization protocols
	D.3.A	Corrosivity
	D.3.B	Irritation potential
	D.3.C	Sensitization potential
Summary:	ICCVAM was established by the Director of the National Institute of Environmental Health Sciences (NIEHS) to implement NIEHS directives to develop and validate new test methods and to establish criteria and processes for the validation and regulatory acceptance of toxicological testing methods. To date, the following dermal assays and associated documents were submitted to ICCVAM for review and evaluation:	

- Dermal corrosivity and irritation assays: Corrositex® Assay; EpiDerm™, EpiSkin™, and Rat Skin Transcutaneous Electrical Resistance (TER) Assay
- Murine Local Lymph Node Assay (LLNA)—a test method for assessing the allergic contact dermatitis potential of chemicals/compounds

Article ID:	**111**
Citation:	International Brotherhood of Teamsters [2006]. [www.teamster.org/].
Resource type:	Web site
Educational materials:	No
Number of references:	
Industries/occupations:	General—overview
Specific process:	
Chemical:	General—overview, plastics and resins, solvents
Specific chemicals:	Diisocyanates
Mixtures:	No

Indexed Dermal Bibliography

Appendix A: Full Resource Citations and Summaries

Audience:	General
Topics addressed:	A Overview
	A.1 Occurrence of skin exposures in the workplace
	A.2 Health hazards resulting from skin exposure to chemicals
	B Exposure characterization
	B.1 Job/tasks, industries/processes, or chemicals associated with skin exposures
	E Risk management
	E.1 Overview of skin exposure control options
Summary:	The International Brotherhood of Teamsters' Safety and Health Department has a Web page on the Teamsters' Web site that contains information on health and safety issues associated with Teamster work activities. A variety of fact sheets in the health and safety section of the Teamster Web site address dermal exposure related issues, including:

- Dermatitis
- Diisocyanates
- Solvents
- General requirements for sanitation

Article ID:	112
Citation:	International Labour Organization (ILO), [2005]. [www.ilo.org/].
Resource type:	Web site
Educational materials:	Yes
Number of references:	
Industries/occupations:	General—overview, Agricultural, Beauty/Cosmetology, Cleaning/Janitorial/Maid, Construction, Forestry/Fisheries, Manufacturing—Chemical, Medical Services, Mining, Service—Food, Service—Medical, Transportation/Communications/Utility
Specific process:	
Chemical:	General—overview, abrasives, cleaning agents, coolants, corrosives, fiberglass and other fibers, food products, hand cleansers, heavy metals/inorganic compounds, latex, nanoparticles, organic dyes, particulates, pesticides, petroleum products & lubricants, plastics and resins, PAHs, PCBs, rubber additives, soaps and detergents, solvents
Specific chemicals:	

Appendix A: Full Resource Citations and Summaries

Mixtures:	Yes	
Audience:	Professional	
Topics addressed:	A	Overview
	A.1	Occurrence of skin exposures in the workplace
	A.2	Health hazards resulting from skin exposure to chemicals
	A.3	Investigation, intervention, and control of occupational skin exposures
	A.4	Skin physiology and function as barriers to chemical insults
	A.5	Dermal regulations and skin notations
	B	Surveillance and clinical aspects
	B.1	Surveillance study reporting incidences of occupational skin exposures
	B.1.A	Skin exposure major focus
	B.1.B	Skin exposure minor focus
	B.2	Loss of workdays and impact on productivity
	C	Exposure characterization
	C.1	Workplace factors associated with harmful skin exposures
	C.2	Description of factors influencing exposure conditions
	C.2.A	Exposure intensity/frequency/duration
	C.2.B	Exposure concentration
	C.2.C	Skin area affected
	D	Hazard identification
	D.1	Potential health effects resulting from specific chemicals
	D.1.A	Irritant contact dermatitis
	D.1.B	Allergic contact dermatitis/sensitization
	D.1.C	Systemic toxicity
	D.1.E	Contribution to overall exposure
	F	Risk management
	F.1	Exposure control strategies
	F.1.A	Substitution
	F.1.B	Engineering controls
	F.1.C	Work practice/Administrative controls
	F.1.D	PPE and PPE rules

Indexed Dermal Bibliography

Appendix A: Full Resource Citations and Summaries

F.1.E Skin management, barrier creams, moisturizers, cleansers, and rubs

Summary: The ILO is a specialized agency of the United Nations that promotes internationally recognized human and labor rights. Among other things, it provides technical assistance in the field of occupational safety and health. A variety occupational safety and health resources containing information on dermal exposure are available through this Web site. Key resources available through ILO include:

- International Occupational Safety and Health Information Centre (CIS): CIS was established in 1959 with the aim of facilitating the exchange of information about occupational safety and health being published around the world, regardless of the format or the language. A variety databases and services that contain dermal exposure information can be accessed through this center (http://www.ilo.org/public/english/protection/safework/cis/products/dbs.htm), including some of those listed below.

- *ILO Encyclopedia of Occupational Health and Safety*: This searchable web version of the encyclopedia includes sections on occupational skin diseases, occupational contact dermatitis, and the prevention of occupational dermatoses. The Encyclopedia can be searched by chemical, industry, and occupation or for potential dermal hazards within each of the previously described categories.

- International Chemical Safety Cards: This searchable database summarizes essential health and safety information on chemicals used by workers and employers in factories, agriculture, construction, and other work places. Information on skin exposure potential is included. This information is available in a wide variety of languages.

- International Risk Phrases Definitions: Risk phrases used by countries in the European Union, including phrases used to classify dermal exposure risks.

- International Hazard Datasheets on Occupations: the International Hazard Datasheets on Occupations are a multipurpose information resource containing information on hazards, risks, and guidelines for prevention related to specific occupations. Dermal hazards associated with listed occupations, when present, are given.

Appendix A: Full Resource Citations and Summaries

Article ID:	113
Citation:	International Safety Equipment Association (ISEA) [2005]. American national standard for hand protection selection criteria. Arlington, VA: ISEA, Report #ANSI/ISEA 105-2005.
Resource type:	Guideline
Educational materials:	No
Number of references:	
Industries/occupations:	General—overview
Specific process:	
Chemical:	General—overview
Specific chemicals:	
Mixtures:	No
Audience:	Professional
Topics addressed:	F Risk management
	F.1 Exposure control strategies
	F.1.D PPE and PPE rules
Summary:	This publication addresses the classification and testing of hand protection for specific performance properties related to chemical and industrial applications. Hand protection includes gloves, mittens, partial gloves, or other items covering the hand or a portion of the hand intended to provide protection against or resistance to a specific hazard. Information can be downloaded from the International Safety Equipment Association's Web site (http://www.safetyequipment.org/glovestd.htm)

Article ID:	114
Citation:	Institute of Medicine (IOM) [2004]. CEFIC workshop on methods to determine dermal permeation for human risk assessment. Riccarton, Edinburgh, UK: IOM, Research Report TM/04/07.
Resource type:	Technical publication/report
Educational materials:	No
Number of references:	40
Industries/occupations:	
Specific process:	
Chemical:	
Specific chemicals:	
Mixtures:	No

Appendix A: Full Resource Citations and Summaries

Audience:	Professional	
Topics addressed:	A	Overview
	A.4	Skin physiology and function as barriers to chemical insults
	C	Exposure characterization
	C.2	Description of factors influencing exposure conditions
	C.2.E	Uptake
	C.5	Exposure modeling
	D	Hazard identification
	D.3	Characterization protocols
	D.3.E	Measurement of skin permeation rates and reservoir effects
	D.3.F	QSAR—development, validation, and application
	E	Risk assessment
	E.2	Example of risk assessments

Summary: This workshop's aim was to develop recommendations for methods to determine dermal permeation rates for use in human risk assessments. This was proposed to be done within the context of the possible future regulatory framework for chemical risk assessment (REACH). The main outcomes of the meeting were:

- A definition of a standardized protocol for an *in vitro* method for measuring dermal absorption of industrial chemicals after infinite and finite doses, to be used to produce data for the development of predictive relationships.
- Recommendations on the existing status and reliability of QSAR data.
- Recommendations on the role of model predictions in generating absorption data for risk assessment.
- Recommendations for a strategy for using measurements and predictions of dermal permeation to meet the requirements of REACH.
- Suggestions on the steps that will be needed to develop this strategy.

Article ID:	**115**
Citation:	Kalnas J, Teitelbaum DT [2000]. Dermal absorption of benzene: Implications for work practices and regulations. Int J Occup Environ Health 6(2):114–21.

Appendix A: Full Resource Citations and Summaries

Resource type:	Journal article—review, meta-analysis
Educational materials:	No
Number of references:	49
Industries/occupations:	
Specific process:	
Chemical:	solvents
Specific chemicals:	benzene
Mixtures:	No
Audience:	Professional

Topics addressed:

	A	Overview
	A.1	Occurrence of skin exposures in the workplace
	A.5	Dermal regulations and skin notations
	C	Exposure characterization
	C.2	Description of factors influencing exposure conditions
	C.2.E	Uptake
	C.5	Exposure modeling
	E	Risk assessment
	E.1	Guidelines for risk assessment or analysis
	E.1.B	Systemic health effects
	E.2	Example of risk assessments

Summary: This paper provides an overview of occupational dermal exposure to benzene. Topics discussed include estimates of the amount of benzene absorbed through the skin and the increased likelihood of developing leukemia at low exposure levels, the development of permissible exposure limits for benzene, and proposed exposure limits from the NIOSH and ACGIH.

Article ID:	**116**
Citation:	Kampf G, Loffler H [2003]. Dermatological aspects of a successful introduction and continuation of alcohol-based hand rubs for hygienic hand disinfection. J Hosp Infect 55(1).
Resource type:	Journal article—review, meta-analysis
Educational materials:	No
Number of references:	61
Industries/occupations:	Service—Medical
Specific process:	
Chemical:	cleaning agents, hand cleansers, soaps and detergents

Appendix A: Full Resource Citations and Summaries

Specific chemicals:		
Mixtures:	No	
Audience:	Professional	
Topics addressed:	D	Hazard identification
	D.1	Potential health effects resulting from specific chemicals
	D.1.A	Irritant contact dermatitis
	F	Risk management
	F.1	Exposure control strategies
	F.1.E	Skin management, barrier creams, moisturizers, cleansers, and rubs
Summary:	This review of alcohol-based hand rubs in hospitals found that after years of hand washing, 30% of healthcare workers incur occupational hand dermatitis, mostly contact dermatitis (allergic reactions were rare). Steps to prevent irritant contact dermatitis are also presented.	

Article ID:	**117**
Citation:	Kanerva L [2000]. Handbook of occupational dermatology. Berlin, NY: Springer.
Resource type:	Book/monograph, whole
Educational materials:	No
Number of references:	8205
Industries/occupations:	Beauty/Cosmetology, Construction, Manufacturing—Other
Specific process:	Aircraft Industry Workers, Air Hammer Operators, Aromatherapists, Asphalt Workers (Paving), Automobile Mechanics, Bakers, Barbers and Hairdressers, Bartenders, Bath Attendants, Batik Manufacturing Workers, Battery Makers, Beekeepers, Biotechnical Industry Workers, Boat Builders, Brake-Lining Workers, Butchers and Slaughterhouse Workers, Cabinet Makers, Candle Makers, Confectionery and Candy Makers, Carpenters Car Industry, Cement Workers, Ceramic and Pottery Workers, Cheese Makers, Chemists, Child Daycare Workers, Cigarette and Cigar Makers, Construction Workers, Cosmetologists, Dental Personnel, Detergent Workers, Divers, Electron Microscopy Workers, Electronic Workers, Electroplaters, Embalmers, Engravers, Farmers and Farm Workers, Floor Layers, Florists, Forestry Workers, Foundry Workers, Fur Farming and the Fur Industry, Furniture Manufacture, Gardeners, Glass Workers, Grinders and Brazers of Hard Metal, Hairdressers, Health Care Workers,

Appendix A: Full Resource Citations and Summaries

	Highway Construction Workers, Histology Technicians, House Workers, Insulation Workers, Jewelers, Laboratory Technicians, Leather Industry, Locksmiths, Machinists, Masseurs, Mechanics, Metal Industry, Metal Polishers, Military Personnel, Mining (Tunneling) Workers, Musicians, Office Workers, Oil-Rig Workers, Operating-Room Staff, Painters, Lacquerers and Varnishers, Paper and Pulp Workers, Pharmaceutical and Cosmetic Industries Workers, Photographers and Other Photo-Lab Workers, Poultry Processors, Pitch Workers, Plumbers and Pipe Fitters, Printers and Lithographers, Professional Sports: Skin Disorders in Athletes, Railroad Shop Workers, Reindeer Herders, Roofers, Shoe Manufacturers and Repairers, Silk-Screen Workers, Stonemasons, Sugar Artists, Swimming Pool Workers, Tattoo Artists, Textile Workers, Veterinary Surgeons, Welders, Winemakers
Chemical:	Hand cleansers, heavy metals/inorganic compounds, latex, pesticides, plastics and resins, rubber additives, solvents, other: disinfectants, formaldehyde, pharmaceuticals, fragrances, enzymes, cement, textiles, leather, adhesives, electronics, paints, polymers, cutting fluids, rubber, plants, spices, and woods.
Specific chemicals:	
Mixtures:	No
Audience:	Professional
Topics addressed:	D Hazard identification
	D.1 Potential health effects resulting from specific chemicals
	D.1.A Irritant contact dermatitis
	D.1.B Allergic contact dermatitis/sensitization
	D.1.C Systemic toxicity
	F Risk management
	F.1 Exposure control strategies
	F.1.C Work practice/Administrative controls
	F.1.D PPE and PPE rules
	F.1.E Skin management, barrier creams, moisturizers, cleansers, and rubs
Summary:	This comprehensive handbook is comprised of dozens of separate articles on occupation skin exposures and is designed to provide information to healthcare professionals for dealing with patients. Diseases covered include: allergic and irritant dermatitis (AID), contact urticaria, photodermatoses, infectious diseases, skin tumors, systematic reactions due to percutaneous

Appendix A: Full Resource Citations and Summaries

absorption, predisposed diseases, and occupational skin problems.

The handbook is divided into four sections.

Part 1—Epidemiology, Treatment, and Prognosis (57 articles).

Part 2—Substances and Products (articles on chemical substances) (36 articles)

Part 3—Job Descriptions with their Irritants and Allergens (94 articles)

Part 4—Chemistry and Concentrations of Patch test Allergens (3 articles)

Article ID:	118
Citation:	Kimber, I [1996]. The role of the skin in the development of chemical respiratory hypersensitivity. Toxicol Lett 86(2–3):89–92.
Resource type:	Journal article—review, meta-analysis
Educational materials:	No
Number of references:	29
Industries/occupations:	
Specific process:	
Chemical:	
Specific chemicals:	
Mixtures:	No
Audience:	Professional
Topics addressed:	A Overview
	A.2 Health hazards resulting from skin exposure to chemicals
	D Hazard identification
	D.1 Potential health effects resulting from specific chemicals
	D.1.B Allergic contact dermatitis/sensitization
	D.1.C Systemic toxicity
	D.1.D Other health effects
Summary:	This paper examines the mechanisms relevant to the stimulation of respiratory sensitization following cutaneous exposure to chemical allergens and implications for the prevention of occupational asthma.

Appendix A: Full Resource Citations and Summaries

Article ID:	**119**
Citation:	Kimber I, Pichowski JS, Betts CJ, Cumberbatch M, Basketter DA, Dearman RJ [2001]. Alternative approaches to the identification and characterization of chemical allergens. Toxicol *In Vitro* 15(4–5):307–12.
Resource type:	Journal article—review, meta-analysis
Educational materials:	No
Number of references:	29
Industries/occupations:	
Specific process:	
Chemical:	
Specific chemicals:	
Mixtures:	No
Audience:	Professional
Topics addressed:	D Hazard identification
	D.1 Potential health effects resulting from specific chemicals
	D.1.B Allergic contact dermatitis/sensitization
	D.3 Characterization protocols
	D.3.B Irritation potential
	D.3.C Sensitization potential
Summary:	This paper describes some of the general requirements of *in vitro* test methods for skin sensitization and progress that has been made in developing suitable approaches with particular emphasis on the utility of dendritic cell culture systems.

Article ID:	**120**
Citation:	Kissel J, Fenske R [2000]. Improved estimation of dermal pesticide dose to agricultural workers upon reentry. Appl Occup Environ Hyg 15(3):284–90.
Resource type:	Journal article—review, meta-analysis
Educational materials:	No
Number of references:	19
Industries/occupations:	Agricultural
Specific process:	
Chemical:	
Specific chemicals:	

Appendix A: Full Resource Citations and Summaries

Mixtures:	No
Audience:	Professional
Topics addressed:	C Exposure characterization
	C.2 Description of factors influencing exposure conditions
	C.2.E Uptake
	C.5 Exposure modeling
Summary:	This article presents a method for agricultural worker dermal dose estimation which accounts for the effect of delay in post-shift washing on dose.

Article ID:	**121**
Citation:	Kissel JC, Richter KY, Fenske RA [1996]. Factors affecting soil adherence to skin in hand-press trials. Bull Environ Contam Toxicol 56(5):722–28.
Resource type:	Journal article—review, meta-analysis
Educational materials:	No
Number of references:	12
Industries/occupations:	
Specific process:	
Chemical:	
Specific chemicals:	
Mixtures:	No
Audience:	Professional
Topics addressed:	C Exposure characterization
	C.2 Description of factors influencing exposure conditions
	C.2.C Skin area affected
Summary:	This paper compares three approaches for estimating soil adherence to the skin for use in dermal risk estimates (listed below). Each approach offers information of value. Laboratory studies provide an opportunity for systematic examination of the possible effects of soil characteristics on adherence.

1. Laboratory studies using artificial loading scenarios (Que Hee, 1985; Driver, 1989; Sheppard and Evenden, 1992).

2. Pb exposure studies reporting Pb concentrations in soil and dust (Roels, 1980; Charney, 1980; Gallacher, 1984; Duggan, 1985).

Appendix A: Full Resource Citations and Summaries

3. Direct field measurement using gravimetric methods (Lepow, 1975).

Article ID:	**122**
Citation:	Klingner TD, Boeniger [2002] MF. A critique of assumptions about selecting chemical-resistant gloves: A case for workplace evaluation of glove efficacy. Appl Occup Environ Hyg *17*(5):360–67.
Resource type:	Journal article—review, meta-analysis
Educational materials:	No
Number of references:	48
Industries/occupations:	
Specific process:	
Chemical:	
Specific chemicals:	
Mixtures:	Yes
Audience:	Professional
Topics addressed:	A Overview
	A.5 Dermal regulations and skin notations
	B Surveillance and clinical aspects
	B.1 Surveillance study reporting incidences of occupational skin exposures
	B.1.A Skin exposure major focus
	D Hazard identification
	D.1 Potential health effects resulting from specific chemicals
	D.1.A Irritant contact dermatitis
	D.1.B Allergic contact dermatitis/sensitization
	D.1.C Systemic toxicity
	F Risk management
	F.1 Exposure control strategies
	F.1.D PPE and PPE rules
Summary:	Those selecting gloves should not rely only upon the manufacturers' laboratory-generated chemical permeation data, for this data may not reflect conditions in the actual workplace (e.g., elevated temperature, flexing, pressure, and product variation between suppliers). This article presents glove selection criteria and recommends dermal monitoring to evaluate glove performance under actual use conditions.

Appendix A: Full Resource Citations and Summaries

Article ID:	123
Citation:	Klotz A, Veeger M, Rocher W [2003]. Skin cleansers for occupational use: testing the skin compatibility of different formulations. Int Arch Occup Environ Health 76(5):367.
Resource type:	Journal article—primary
Educational materials:	No
Number of references:	17
Industries/occupations:	
Specific process:	
Chemical:	Abrasives, hand cleansers, solvents
Specific chemicals:	
Mixtures:	No
Audience:	Professional
Topics addressed:	A Overview
	A.4 Skin physiology and function as barriers to chemical insults
	F Risk management
	F.1 Exposure control strategies
	F.1.E Skin management, barrier creams, moisturizers, cleansers, and rubs
Summary:	This article presents an overview of skin-cleansing products and their ingredients and discusses skin compatibility and cleansing effectiveness. The authors advocate a range of skin cleansers depending upon the degree of contamination. They also recommend avoiding solvents and abrasives to prevent occupational dermatitis, and stress the importance of worker education.

Article ID:	124
Citation:	Koch P [2001]. Occupational contact dermatitis: recognition and management. Am J Clin Dermatol 2(6):353–65.
Resource type:	Journal article—review, meta-analysis
Educational materials:	No
Number of references:	84
Industries/occupations:	Beauty/Cosmetology, Construction, Manufacturing—Other, Service—Food, Service—Medical
Specific process:	Hairdressers, dental laboratory technicians, healthcare workers, metal workers, leather and shoe workers, bakers, caterers, confectioners, and cooks

Appendix A: Full Resource Citations and Summaries

Chemical:	Latex, organic dyes, pesticides, rubber additives, solvents, other: concrete, glues, leather	
Specific chemicals:	Use, concentration, associated industry, and health effects are presented for dozens of chemicals.	
Mixtures:	No	
Audience:	Professional	
Topics addressed:	A	Overview
	A.2	Health hazards resulting from skin exposure to chemicals
	C	Exposure characterization
	C.2	Description of factors influencing exposure conditions
	C.2.B	Exposure concentration
	D	Hazard identification
	D.1	Potential health effects resulting from specific chemicals
	D.1.A	Irritant contact dermatitis
Summary:	This paper primarily focuses on eight broad occupational categories at risk for occupational contact dermatitis (OCD) and dozens of their associated chemical irritants and sensitizers. The categories are hairdressers, dental laboratory technicians, healthcare workers, construction industry workers, metal workers, leather and shoe workers, florists and gardeners, and food service workers (bakers, caterers, confectioners, and cooks).	

Article ID:	**125**	
Citation:	Kresken J, Klotz A [2003]. Occupational skin-protection products—a review. Int Arch Occup Environ Health 76(5):355–58.	
Resource type:	Journal article—review, meta-analysis	
Educational materials:	No	
Number of references:	34	
Industries/occupations:		
Specific process:		
Chemical:	Hand cleansers, water	
Specific chemicals:		
Mixtures:	No	
Audience:	Professional	
Topics addressed:	D	Hazard identification

Indexed Dermal Bibliography **197**

Appendix A: Full Resource Citations and Summaries

	D.1	Potential health effects resulting from specific chemicals
	D.1.A	Irritant contact dermatitis
	D.1.B	Allergic contact dermatitis/sensitization
	F	Risk management
	F.1	Exposure control strategies
	F.1.E	Skin management, barrier creams, moisturizers, cleansers, and rubs
Summary:	This paper evaluates the use of skin protection products, including barrier creams, in preventing occupational dermal exposures. They conclude barrier creams do not replace PPE and should only be used against low-grade irritants such as water, detergents, and cutting fluids.	

Article ID:	**126**	
Citation:	Kromhout H, Vermeulen R [2001]. Temporal, personal and spatial variability in dermal exposure. Ann Occup Hyg 45(4):257–73.	
Resource type:	Journal article—review, meta-analysis	
Educational materials:	No	
Number of references:	28	
Industries/occupations:	Agricultural, Construction, Manufacturing	
Specific process:	Rubber manufacturing, Asphalt paving, Coke production	
Chemical:	Pesticides, PAHs, paint	
Specific chemicals:		
Mixtures:	Yes	
Audience:	Professional	
Topics addressed:	C	Exposure characterization
	C.1	Workplace factors associated with harmful skin exposures
	C.2	Description of factors influencing exposure conditions
	C.2.A	Exposure intensity/frequency/duration
	C.2.C	Skin area affected
	C.5	Exposure modeling
	E	Risk assessment
	E.2	Example of risk assessments
Summary:	A database of dermal exposure measurements (DERMDAT) comprising data from 20 surveys was created	

from agricultural and industrial workers containing 6,400 observations. Analyses of variability showed median values of the total, within-, and between-worker geometric standard deviations to be similar to that published previously for respiratory exposure.

Article ID:	**127**
Citation:	Kutting B, Drexler H [2003]. Effectiveness of skin protection creams as a preventive measure in occupational dermatitis: a critical update according to criteria of evidence-based medicine. Int Arch Occup Environ Health 76(4):253–59.
Resource type:	Journal article—review, meta-analysis
Educational materials:	No
Number of references:	63
Industries/occupations:	
Specific process:	
Chemical:	
Specific chemicals:	
Mixtures:	No
Audience:	Professional
Topics addressed:	D Hazard identification
	D.1 Potential health effects resulting from specific chemicals
	D.1.A Irritant contact dermatitis
	F Risk management
	F.1 Exposure control strategies
	F.1.E Skin management, barrier creams, moisturizers, cleansers, and rubs
Summary:	This paper reviews the literature to answer the questions: (1) Can a skincare regimen effectively reduce or eliminate work-related poor skin conditions? (2) Do protective creams prevent harmful substances from penetrating and adhering to the skin? (3) Is the differentiation between preexposure and postexposure products justified by reliable data? The authors also address the merit of the traditional three-step skin protection program: skin protection before work, cleaning, and skin care after work. This paper concludes that not enough data have been accumulated to prove the benefit of skin protection measures under real workplace conditions.

Appendix A: Full Resource Citations and Summaries

Article ID:	128
Citation:	Leggat PA, Kedjarune U [2003]. Toxicity of methyl methacrylate in dentistry. Int Dent J 53(3):126–31.
Resource type:	Journal article—review, meta-analysis
Educational materials:	No
Number of references:	50
Industries/occupations:	Service—Medical, Dentistry
Specific process:	
Chemical:	Plastics and resins
Specific chemicals:	Methyl methacrylate (MMA)
Mixtures:	No
Audience:	Professional
Topics addressed:	B Surveillance and clinical aspects
	B.1 Surveillance study reporting incidences of occupational skin exposures
	B.1.B Skin exposure minor focus
	D Hazard identification
	D.1 Potential health effects resulting from specific chemicals
	D.1.C Systemic toxicity
	F Risk management
	F.1 Exposure control strategies
	F.1.B Engineering controls
	F.1.C Work practice/Administrative controls
	F.1.D PPE and PPE rules
Summary:	This paper presents health effects associated with exposure to MMA in dentistry. It includes a discussion of control strategies to use to reduce exposure to MMA.

Article ID:	129
Citation:	Finnish Institute of Occupational Health (FIOH) [2001]. Epidemiology of skin and respiratory diseases among hairdressers. Helsinki, Finland: FIOH
Resource type:	Technical publication/report
Educational materials:	No
Number of references:	298
Industries/occupations:	Beauty/Cosmetology
Specific process:	Hairdressers, salons

Appendix A: Full Resource Citations and Summaries

Chemical:	Organic dyes, soaps and detergents	
Specific chemicals:	Ammonium persulfate	
Mixtures:	No	
Audience:	Professional	
Topics addressed:	A	Overview
	A.1	Occurrence of skin exposures in the workplace
	A.2	Health hazards resulting from skin exposure to chemicals
	B	Surveillance and clinical aspects
	B.1	Surveillance study reporting incidences of occupational skin exposures
	B.1.A	Skin exposure major focus
	B.3	Surveillance study protocols/procedures for gathering data
	C	Exposure characterization
	C.1	Workplace factors associated with harmful skin exposures
	C.4	Direct methods to measure exposure
	C.4.B	Skin
	D	Hazard identification
	D.1	Potential health effects resulting from specific chemicals
	D.1.A	Irritant contact dermatitis
	D.1.B	Allergic contact dermatitis/sensitization
	D.1.D	Other health effects
	F	Risk management
	F.1	Exposure control strategies
	F.1.B	Engineering controls
Summary:	This paper includes the following: Section 1 presents introductory material on dermal and respiratory exposure among hairdressers (15 pages). Section 2 presents findings of a literature review (27 pages). Section 3 to 7 present methods, results, discussion, and conclusions regarding 20 Finnish salons (50 pages). Section 8 is the reference list (32 pages).	

In addition to an extensive literature review, this paper presents five epidemiologic studies of skin and respiratory disorders among hairdressers. Study I focuses on the working conditions in salons and the perceived health of the hairdressers. Studies II, III, and IV focus on the prevalence, incidence, and risk of skin and respiratory symptoms and diseases among hairdressers. Study

Appendix A: Full Resource Citations and Summaries

V focuses on the risks of and causes for leaving the profession. Environmental data were collected at 20 Finnish salons. Health data were collected by questionnaire, phone interviews, and medical examinations. The hairdressing salons meet Finnish indoor air criteria, but high peak concentrations of certain chemicals, including ammonium persulfate, were found to cause skin and respiratory diseases. Hairdressers incur an increased incidence of asthma and chronic bronchitis. Local exhaust ventilation was recommended.

Ammonium persulfate is used as a polymerization initiator in polymer chemistry, as an etchant and cleaner in manufacture of printed circuit boards, as a booster in hair bleaching formulations in cosmetics, and as a secondary oil recovery system by acting as a polymerization initiator and a gel breaker.

Article ID:	130
Citation:	Leung HW, Paustenbach DJ [1994]. Techniques for estimating the percutaneous absorption of chemicals due to occupational and environmental exposure. Appl Occup Environ Hyg 9(3):187-97.
Resource type:	Journal article—review, meta-analysis
Educational materials:	No
Number of references:	92
Industries/occupations:	
Specific process:	
Chemical:	
Specific chemicals:	
Mixtures:	No
Audience:	Professional
Topics addressed:	C Exposure characterization
	C.2 Description of factors influencing exposure conditions
	C.2.A Exposure intensity/frequency/duration
	C.2.B Exposure concentration
	C.2.C Skin area affected
	C.2.E Uptake
	C.4 Direct methods to measure exposure
	C.4.A Surfaces
	C.5 Exposure modeling

Appendix A: Full Resource Citations and Summaries

	D	Hazard identification
	D.3	Characterization protocols
	D.3.E	Measurement of skin permeation rates and reservoir effects
Summary:		This paper reviewed techniques for estimating the percutaneous absorption of chemicals following occupational exposure. It discusses factors influencing percutaneous absorption including number of exposures, nature of broken skin, exposure site, chemical uptake, and skin surface area. The latter was considered to be the most important factor. Discusses absorption studies, modeling, calculating exposure, and the interpretation of wipe sample data.

Article ID:	**131**	
Citation:	Liu Y, Bello D, Sparer JA, Stowe MH, Gore RJ, Woskie SR, Cullen MR, Redlich CA [2007]. Skin exposure to aliphatic polyisocyanates in the auto body repair and refinishing industry: A qualitative assessment. Ann Occup Hyg 51(5):429–39.	
Resource type:	Journal article—primary	
Educational materials:	No	
Number of references:	22	
Industries/occupations:	Manufacturing—Automotive auto body repair, refinishing	
Specific process:	Autobody workers	
Chemical:	Paint	
Specific chemicals:	Aliphatic polyisocyanate, hexamethylene diisocyanate (HDI), isophorone diisocyanate (IPDI)	
Mixtures:	No	
Audience:	Professional	
Topics addressed:	A	Overview
	A.1	Occurrence of skin exposures in the workplace
	B	Surveillance and clinical aspects
	B.1	Surveillance study reporting incidences of occupational skin exposures
	B.1.A	Skin exposure major focus
	C	Exposure characterization
	C.2	Description of factors influencing exposure conditions
	C.2.B	Exposure concentration
	C.2.C	Skin area affected

Appendix A: Full Resource Citations and Summaries

C.2.E	Uptake
C.4	Direct methods to measure exposure
C.4.A	Surfaces
C.4.B	Skin
D	Hazard identification
D.3	Characterization protocols
D.3.E	Measurement of skin permeation rates and reservoir effects
D.4	Other
E	Risk assessment
E.3	Other
F	Risk management
F.1	Exposure control strategies
F.1.D	PPE and PPE rules

Summary: This study evaluated aliphatic isocyanate skin exposure among auto body shop workers. Also evaluated was the effectiveness of gloves and other PPE.

Article ID:	132
Citation:	Lowney YW, Ruby MV, Wester RC, Schoof RA, Holm S E, Hui XY, Barbadillo S, Maibach HI [2005]. Percutaneous absorption of arsenic from environmental media. Toxicol Ind Health 21(1–2):1–14.
Resource type:	Journal article—review, meta-analysis
Educational materials:	No
Number of references:	30
Industries/occupations:	
Specific process:	
Chemical:	
Specific chemicals:	Arsenic
Mixtures:	No
Audience:	Professional
Topics addressed:	

C	Exposure characterization
C.2	Description of factors influencing exposure conditions
C.2.B	Exposure concentration
C.2.E	Uptake

Appendix A: Full Resource Citations and Summaries

Summary:	This paper addresses what is known about percutaneous absorption of arsenic based on studies of rhesus monkeys and offers study design considerations including particle size, application rates, means of ensuring skin contact, and appropriate statistical evaluation of the data. The authors conclude that there are likely to be many site- or sample-specific factors that control the absorption of arsenic, and matrix-specific analyses may be required to understand the degree of percutaneous absorption.
Article ID:	**133**
Citation:	Lushniak BD [1995]. The epidemiology of occupational contact dermatitis. Dermatol Clin *13*(3):671–80.
Resource type:	Journal article—review, meta-analysis
Educational materials:	No
Number of references:	66
Industries/occupations:	Agricultural, Construction, Forestry/Fisheries, Manufacturing—Chemical, Manufacturing—Other, Mining, Service—Food, Service—Medical, Service—Transportation/Communications/Utility, Trade, finance/insurance/realty, meat products, leather, motorvehicles
Specific process:	
Chemical:	
Specific chemicals:	
Mixtures:	No
Audience:	Professional
Topics addressed:	B Surveillance and clinical aspects
	B.1 Surveillance study reporting incidences of occupational skin exposures
	B.1.A Skin exposure major focus
	B.2 Loss of workdays and impact on productivity
	B.3 Surveillance study protocols/procedures for gathering data
	D Hazard identification
	D.1 Potential health effects resulting from specific chemicals
	D.1.A Irritant contact dermatitis
Summary:	This article reviews occupational contact dermatitis epidemiologic data sources for important information on prevalence, diagnosis, public health importance, risk factors, etiologic agents, prognosis, and preventive

Appendix A: Full Resource Citations and Summaries

measures. It also provides incidences for different occupational groups.

Article ID:	**134**
Citation:	Lushniak BD [2003]. The importance of occupational skin diseases in the United States. Int Arch Occup Environ Health 76(5):325–30.
Resource type:	Journal article—review, meta-analysis
Educational materials:	No
Number of references:	22
Industries/occupations:	General—overview
Specific process:	Provides data by major occupational categories
Chemical:	
Specific chemicals:	
Mixtures:	No
Audience:	Professional
Topics addressed:	A Overview
	A.1 Occurrence of skin exposures in the workplace
	B Surveillance and clinical aspects
	B.1 Surveillance study reporting incidences of occupational skin exposures
	B.1.A Skin exposure major focus
	B.2 Loss of workdays and impact on productivity
Summary:	This epidemiological study presents occupational skin disease and disorder surveillance data for the U.S. and three states (OH, OR, and WA). It describes trends, data by occupation, lost time, and other data.

Article ID:	**135**
Citation:	Lushniak BD [2004]. Occupational contact dermatitis. Dermatol Ther 17(3):272–77.
Resource type:	Journal article—review, meta-analysis
Educational materials:	No
Number of references:	32
Industries/occupations:	General—overview, Agricultural, Construction, Forestry/Fisheries, Manufacturing—Chemical, Mining, Service—Medical, Transportation/Communications/Utility
Specific process:	

Appendix A: Full Resource Citations and Summaries

Chemical:		
Specific chemicals:		
Mixtures:	No	
Audience:	Professional	
Topics addressed:	A	Overview
	A.1	Occurrence of skin exposures in the workplace
	A.2	Health hazards resulting from skin exposure to chemicals
	A.3	Investigation, intervention, and control of occupational skin exposures
	B	Surveillance and clinical aspects
	B.1	Surveillance study reporting incidences of occupational skin exposures
	B.1.A	Skin exposure major focus
	B.2	Loss of workdays and impact on productivity
	C	Exposure characterization
	C.1	Workplace factors associated with harmful skin exposures
	D	Hazard identification
	D.1	Potential health effects resulting from specific chemicals
	D.1.A	Irritant contact dermatitis
	D.1.B	Allergic contact dermatitis/sensitization
	D.1.D	Other health effects
	F	Risk management
	F.1	Exposure control strategies
	F.1.A	Substitution
	F.1.D	PPE and PPE rules
Summary:	This paper presents an overview of issues involved in the study of occupational contact dermatitis, including importance, incidence, economic impact, at-risk occupations, diagnosis, and prevention.	

Article ID:	**136**
Citation:	Mansdorf SZ, Henry N III [2003]. Personal protective clothing. In: DiNardi SR, ed. The occupational environment: Its evaluation, control, and management, 2nd ed. Fairfax, VA: American Industrial Hygiene Association.
Resource type:	Book/monograph, chapter

Indexed Dermal Bibliography

Appendix A: Full Resource Citations and Summaries

Educational materials:	No	
Number of references:	42	
Industries/occupations:		
Specific process:		
Chemical:		
Specific chemicals:		
Mixtures:	No	
Audience:	Professional	
Topics addressed:	F	Risk management
	F.1	Exposure control strategies
	F.1.D	PPE and PPE rules
Summary:	Chapter 35, "Personal Protective Clothing," from the AIHA book, *The Occupational Environmental: Its Evaluation Control and Management (the White Book)*, discusses PPE for chemical hazards as well as thermal, mechanical, radiological, and biological hazards. It discusses performance characteristics, ergonomics, cost, maintenance, and training for different types of personal protective equipment used to control dermal hazards.	

Article ID:	**137**	
Citation:	Marks JG [2002]. Contact and occupational dermatology, 3rd ed. St. Louis: Mosby.	
Resource type:	Book/monograph, whole	
Educational materials:	No	
Number of references:		
Industries/occupations:	General—overview, Agricultural, Beauty/Cosmetology, Cleaning/Janitorial/Maid, Construction, Manufacturing—Other, Medical Services	
Specific process:	Electronics workers, Dental workers, Florists, Food service workers, Machinists, Office workers, Photographers, Printers, Textile	
Chemical:	Food products	
Specific chemicals:	Preservatives	
Mixtures:	No	
Audience:	Professional	
Topics addressed:	B	Surveillance and clinical aspects
	B.4	Clinical protocols for recognition of skin exposure health effects
	C	Exposure characterization

Appendix A: Full Resource Citations and Summaries

	C.4	Direct methods to measure exposure
	C.4.B	Skin
	D	Hazard identification
	D.1	Potential health effects resulting from specific chemicals
	D.1.A	Irritant contact dermatitis
	D.1.B	Allergic contact dermatitis/sensitization
	D.3	Characterization protocols
	D.3.C	Sensitization potential

Summary: This book covers the diagnosis and management of suspected contact and occupational dermatitis. The early chapters of this book focus on the nonoccupational aspects of dermatology but include discussions of dermatitis recognition, monitoring, and treatment, as well as health effects from preservatives, vehicles, cosmetics, fragrances, and hair and nail care. The focus is on treatment rather than prevention.

Chapters 12 through 17 cover occupational dermatology: These chapters are:

Ch. 12 Etiology of Occupational Skin Disease Workers

Ch. 13 Evaluation of the Worker in the Office and at the Work Site

Ch. 14 Management of Occupational Dermatitis

Ch. 15 Occupations Commonly Associated With Contact Dermatitis

Ch. 16 Contact Urticaria

Ch. 17 Contact Dermatitis in Children

Article ID:	**138**
Citation:	Marquart H, Maidment S, McClaflin JL, Fehrenbacher MC. Harmonization of future needs for dermal exposure assessment and modeling: A workshop report. Appl Occup Environ Hyg *16*(2):218–27.
Resource type:	Journal article—review, meta-analysis
Educational materials:	No
Number of references:	42
Industries/occupations:	
Specific process:	

Appendix A: Full Resource Citations and Summaries

Chemical:	
Specific chemicals:	
Mixtures:	No
Audience:	Professional
Topics addressed:	A Overview
	A.5 Dermal regulations and skin notations
	C Exposure characterization
	C.2 Description of factors influencing exposure conditions
	C.2.E Uptake
	C.4 Direct methods to measure exposure
	C.4.A Surfaces
	C.4.B Skin
	C.5 Exposure modeling
Summary:	This article is a summary of the 1999 International Symposium on Occupational Exposure Databases and Their Application for the Next Millennium held in London. The workshop was organized in an effort to harmonize future needs in this area. It discusses what is known about methods to measure the amount of contaminant on the skin and surfaces, the amount of contaminant absorbed through the skin, and merits of these approaches. It also discusses what is needed in the field, including raising awareness among occupational health practitioners and creating simple tools for small- and medium-sized businesses to use in risk assessment and management activities.

Article ID:	**139**
Citation:	Marquart J, Brouwer DH, Gijsbers JH, Links IH, Warren N, van Hemmen JJ [2003]. Determinants of dermal exposure relevant for exposure modeling in regulatory risk assessment. Ann Occup Hyg *47*(8):599–607.
Resource type:	Journal article—review, meta-analysis
Educational materials:	No
Number of references:	71
Industries/occupations:	
Specific process:	
Chemical:	
Specific chemicals:	
Mixtures:	No

Appendix A: Full Resource Citations and Summaries

Audience:	Professional	
Topics addressed:	C	Exposure characterization
	C.1	Workplace factors associated with harmful skin exposures
	C.2	Description of factors influencing exposure conditions
	C.2.A	Exposure intensity/frequency/duration
	C.5	Exposure modeling
	E	Risk assessment
	E.1	Guidelines for risk assessment or analysis
	E.1.A	Localized health effects
	E.1.B	Systemic health effects
Summary:	This article is the 3rd article of a six-part series on RISKOFDERM, a tool for conducting risk assessments. The series was published in the Annals of Occupational Hygiene in 2003. The following briefly summarizes each paper in the series:	

1. ID 212—Outlines a "toolkit" for conducting dermal occupational risk assessment.
2. ID 163—Describes the assumptions in the toolkit and describes an approach to exposure assessment used by the toolkit.
3. ID 139—Describes the determinants relevant for dermal exposure models in the scope of regulatory risk assessment.
4. ID 219—Describes how default dermal exposure values can be adjusted for specific work situations.
5. ID 100—Describes the derivation of the toolkit's default task-based dermal exposure values.
6. ID 193—Describes the development of "intrinsic toxicity" (IT) scores used for hazard characterization.

Article ID:	**140**
Citation:	Mathur AK, Khanna SK [2002]. Dermal toxicity due to industrial chemicals. Skin Pharmacol Appl Skin Physiol *15*(3):147–53.
Resource type:	Journal article—review, meta-analysis
Educational materials:	No
Number of references:	21
Industries/occupations:	
Specific process:	

Indexed Dermal Bibliography **211**

Appendix A: Full Resource Citations and Summaries

Chemical:	Cleaning agents, coolants, heavy metals/inorganic compounds, latex, organic dyes, plastics and resins, rubber additives, soaps and detergents, solvents, other: florescent whitening agents, dyes, adhesives, perfume, preservatives
Specific chemicals:	Dozens of specific chemicals are addressed
Mixtures:	No
Audience:	Professional
Topics addressed:	A Overview
	A.4 Skin physiology and functions as a barrier to chemical insults
	D Hazard identification
	D.1 Potential health effects resulting from specific chemicals
	D.1.A Irritant contact dermatitis
	D.1.B Allergic contact dermatitis/sensitization
	D.1.C Systemic toxicity
	D.1.D Other health effects
	F Risk management
	F.1 Exposure control strategies
	F.1.E Skin management, barrier creams, moisturizers, cleansers, and rubs
Summary:	The paper discusses potential health effects from exposure to metals, florescent whiting agents, dyes, adhesives and resins, preservatives and disinfectants, plastics and rubbers, perfume, soaps and detergents, and cutting oils and solvents.

Article ID:	**141**
Citation:	McArthur B [1992]. Dermal measurement and wipe sampling methods: A review. Appl Occup Environ Hyg 7(9):599–606.
Resource type:	Journal article—review, meta-analysis
Educational materials:	No
Number of references:	81
Industries/occupations:	
Specific process:	
Chemical:	
Specific chemicals:	
Mixtures:	No
Audience:	Professional

Topics addressed:	C	Exposure characterization
	C.4	Direct methods to measure exposure
	C.4.A	Surfaces
	C.4.B	Skin
	C.4.C	Biomonitoring
Summary:		This article discusses various methods for directly measuring dermal exposures to hazardous materials on the skin or clothing and on work surfaces, deposited by patches, skin swabs, rinses, and radioactive or fluorescent tracers. This article also discusses biological monitoring (measuring biomarkers for blood, urine, or exhaled air).

Article ID:	**142**	
Citation:		McClean MD, Rinehart RD, Sapkota A, Cavallari JM, Herrick RF [2007]. Dermal exposure and urinary 1-hydroxypyrene among asphalt roofing workers. J Occup Environ Hyg 4(1). (2007): 118-26
Resource type:		Journal article—primary
Educational materials:		No
Number of references:		23
Industries/occupations:		Construction
Specific process:		Asphalt roofing workers
Chemical:		Petroleum products & lubricants, other: polycyclic aromatic compounds (PACs), coal tar pitch
Specific chemicals:		Pyrene, benzoapyrene (BAP)
Mixtures:		No
Audience:		Professional
Topics addressed:	A	Overview
	A.1	Occurrence of skin exposures in the workplace
	C	Exposure characterization
	C.1	Workplace factors associated with harmful skin exposures
	C.2	Description of factors influencing exposure conditions
	C.2.A	Exposure intensity/frequency/duration
	C.2.B	Exposure concentration
	C.2.E	Uptake
	C.4	Direct methods to measure exposure
	C.4.B	Skin

Appendix A: Full Resource Citations and Summaries

	D	Hazard identification
	D.3	Characterization protocols
	D.3.E	Measurement of skin permeation rates and reservoir effects

Summary: This study ascertained determinants of dermal exposure to polycyclic aromatic compounds (PACs) among asphalt roofing workers using dermal patches and urine samples to evaluate the effect of dermal exposure on total absorbed dose. Specific tasks related to roofing included tearing off old roofs, putting down new roofs, and operating the kettle. Results were presented. Dermal exposure was a significant determinant of total absorbed dose.

Article ID:	143
Citation:	McDougal JN, Robinson PJ [2002]. Assessment of dermal absorption and penetration of components of a fuel mixture. Sci Total Environ 288(1–2):23–30.
Resource type:	Journal article—review, meta-analysis
Educational materials:	No
Number of references:	20
Industries/occupations:	
Specific process:	
Chemical:	Petroleum products & lubricants
Specific chemicals:	JP-8 jet fuel, undecane, dodecane, decane, tridecane, tetradecane, methyl naphthalenes, trimethyl benzene, nonane, pentadecane, dimethyl naphthalene, dimethyl benzene (xylene), naphthalene, ethyl benzene, methyl benzene (toluene)
Mixtures:	Yes
Audience:	Professional

Topics addressed:

C	Exposure characterization
C.2	Description of factors influencing exposure conditions
C.2.E	Uptake
E	Risk assessment
E.2	Example of risk assessments

Summary: This article discusses methods for assessing the risks from dermal exposures to complex mixtures, specifically JP-8 jet fuel—a volatile mixture which varies radically in composition depending on the phase of the mixture (vapor, liquid, or aerosol). This article assesses absorption (into the skin) and penetration (through the skin) of components in the mixture and discusses why absorption and penetration can differ. Permeability coefficients for 12 components in JP-8 jet fuel were calculated. The authors suggest that absorption and penetration methodologies similar to those used for JP-8 jet fuel could be used to estimate systemic toxicity of other mixtures.

Article ID:	**144**
Citation:	McDougal JN, Boeniger MF [2002]. Methods for assessing risks of dermal exposures in the workplace. Crit Rev Toxicol *32*(4):291–327.
Resource type:	Journal article—review, meta-analysis
Educational materials:	No
Number of references:	70
Industries/occupations:	
Specific process:	
Chemical:	
Specific chemicals:	
Mixtures:	No
Audience:	Professional
Topics addressed:	A Overview
	A.3 Investigation, intervention, and control of occupational skin exposures
	A.4 Skin physiology and function as barriers to chemical insults
	C Exposure characterization
	C.2 Description of factors influencing exposure conditions
	C.2.E Uptake
	C.5 Exposure modeling
Summary:	This paper provides a comprehensive and comparative analysis of methods used to estimate both the amount of a chemical contacting the skin (external dose) and the amount that reaches internal organs (internal dose). This paper addresses each step in the process, describes the assumptions involved, assesses the model's strengths and weaknesses, and provides recommendations for further research. The paper discusses the following:

- Internal dose assessment
  — Flux and permeability theory

Appendix A: Full Resource Citations and Summaries

- — Calculations based on empirical measurements and fraction absorbed
- — Calculations based on steady-state flux
- — Calculations adjusted for square root of time
- — Calculations based on biologically based models
- — Comparisons with short-term skin penetration data
- Route-to-route extrapolations
  - — Extrapolation factor approach
  - — Biologically based models
- Dermal exposure levels
  - — Skin notation
  - — Banding approach to dermal exposure risks
  - — Dermal occupational exposure levels
  - — Skin absorption time
- Risk characterizations

Article ID:	**145**
Citation:	McDougal JN, Council EA III, Powers BS [2007]. Systemic toxicity from skin exposures (or what happens when you do not decontaminate). J Chem Health Saf 14(4):23–31.
Resource type:	Journal article—review, meta-analysis
Educational materials:	No
Number of references:	92
Industries/occupations:	
Specific process:	
Chemical:	Corrosives, heavy metals/inorganic compounds, pesticides, solvents, other: pharmaceuticals
Specific chemicals:	
Mixtures:	No
Audience:	General
Topics addressed:	A Overview
	E Risk management
	E.4 Guidelines/recommendations for postexposure skin decontamination
Summary:	This resource provides a general overview of the use and misuse of decontamination after dermal exposure. The author points out that some decontamination procedures can make penetration of a chemical through

Appendix A: Full Resource Citations and Summaries

the skin worse. This review evaluates the necessity for decontamination of various chemical classes in the workplace.

Article ID:	**146**
Citation:	Meding B [2000]. Differences between the sexes with regard to work-related skin disease. Contact Dermatitis 43(2):65–71.
Resource type:	Journal article—review, meta-analysis
Educational materials:	No
Number of references:	41
Industries/occupations:	Beauty/Cosmetology, Cleaning/Janitorial/Maid, Service—Food, Service—Medical
Specific process:	Shows high-risk occupations by major group in 1990
Chemical:	Heavy metals/inorganic compounds
Specific chemicals:	Nickel
Mixtures:	No
Audience:	Professional

Topics addressed:		
	B	Surveillance and clinical aspects
	B.1	Surveillance study reporting incidences of occupational skin exposures
	B.1.A	Skin exposure major focus
	D	Hazard identification
	D.1	Potential health effects resulting from specific chemicals
	D.1.A	Irritant contact dermatitis
	D.1.B	Allergic contact dermatitis/sensitization
	F	Risk management
	F.1	Exposure control strategies
	F.1.C	Work practice/Administrative controls
	F.1.D	PPE and PPE rules

Summary:	This paper reviews gender differences in work-related skin disease. Women report skin disease more often than men. They are more often affected than men, and they work in female-dominated occupations (e.g., hairdressing, catering, cleaning, and health-care work) which are more likely to involve wet work. For these occupations, work-related skin disease is common and usually presents as hand eczema, typically, irritant contact dermatitis. Nickel allergy is the most common contact allergy. Control strategy discussion includes a focus on reducing wet exposure.

Indexed Dermal Bibliography

Appendix A: Full Resource Citations and Summaries

Article ID:	**147**
Citation:	Mellstrom GA, Wrangsjo K, Wahlberg JE, Fryklund B [1996]. The value and limitations of protective gloves in medical health service: part III. Dermatol Nurs *8*(5):345–55.
Resource type:	Journal article—review, meta-analysis
Educational materials:	No
Number of references:	14
Industries/occupations:	Service—Medical
Specific process:	
Chemical:	latex
Specific chemicals:	
Mixtures:	No
Audience:	General

Topics addressed:	A	Overview
	A.2	Health hazards resulting from skin exposure to chemicals
	E	Risk management
	E.3	"Best practices"/guidelines/recommendations
	E.3.D	PPE and PPE rules

Summary:	This paper presents the benefits and problems associated with glove protection from dermal exposures. Describes problems of permeability and side effects (latex allergy).

Article ID:	**148**
Citation:	Moss GP, Dearden JC, Patel H, Cronin MT [2002]. Quantitative structure-permeability relationships (QSPRs) for percutaneous absorption. Toxicol *In Vitro 16*(3):299–317.
Resource type:	Journal article—review, meta-analysis
Educational materials:	No
Number of references:	123
Industries/occupations:	
Specific process:	
Chemical:	
Specific chemicals:	
Mixtures:	No
Audience:	Professional

Appendix A: Full Resource Citations and Summaries

Topics addressed:	C	Exposure characterization
	C.2	Description of factors influencing exposure conditions
	C.2.E	Uptake
	D	Hazard identification
	D.3	Characterization protocols
	D.3.E	Measurement of skin permeation rates and reservoir effects
	D.3.F	QSARs—development, validation, and application
Summary:		This article reviews the use and validity of the current state-of-the-art in quantitative structure property relationships (QSPRs) and, more specifically, quantitative structure activity relationship (QSARs) used in modeling the absorption of chemicals through the skin.

Article ID:	**149**
Citation:	Nash JL [2000]. Skin care: starting from scratch. Occupational Hazards *62*(4):53–55.
Resource type:	Magazine article
Educational materials:	Yes
Number of references:	0
Industries/occupations:	
Specific process:	
Chemical:	Other
Specific chemicals:	
Mixtures:	No
Audience:	General

Topics addressed:	A	Overview
	A.1	Occurrence of skin exposures in the workplace
	E	Risk management
	E.1	Overview of skin exposure control options
	E.3	"Best practices"/guidelines/recommendations
	E.3.A	Substitution
	E.3.B	Engineering controls
	E.3.C	Work practice/administration controls
	E.3.D	PPE and PPE rules
	E.3.E	Skin management, barrier creams, moisturizers, cleansers, and rubs

Appendix A: Full Resource Citations and Summaries

Summary:	This paper provides an overview of occupational skin disease, its underreporting, and prevention.
Article ID:	**150**
Citation:	NIOSH [2005]. Recommendations for CPC: A companion to the NIOSH pocket guide to chemical hazards. [www.cdc.gov/niosh/ncpc/ncpc1.html].
Resource type:	Web page
Educational materials:	No
Number of references:	
Industries/occupations:	General—overview
Specific process:	
Chemical:	Abrasives, cleaning agents, coolants, corrosives, fiberglass and other fibers, heavy metals/inorganic compounds, organic dyes, particulates, pesticides, petroleum products & lubricants, plastics and resins, solvents
Specific chemicals:	Includes all chemicals in the NIOSH Pocket Guide to Chemical Hazards (ID 152)
Mixtures:	No
Audience:	General
Topics addressed:	E Risk management
	E.3 "Best practices"/guidelines/recommendations
	E.3.D PPE and PPE rules
Summary:	This Web page provides CPC recommendations for all chemicals listed in the *NIOSH Pocket Guide to Chemical Hazards*, June 1997 Edition (NIOSH Publication No. 97-140). These recommendations are based on another published work, *Quick Selection Guide to Chemical Protective Clothing, Third Edition*, by Krister Forsberg and S.Z. Mansdorf (1997).

The Pocket Guide provides general recommendations in table format for skin protection according to the following designations:

- Prevent skin contact, meaning that there is a dermal hazard potential.
- Frostbite, meaning there is the potential for freezing of the skin from direct contact with the liquified gas through rapid evaporation.
- N.R. means that no recommendation can be made either because the chemical is not a demonstrated dermal hazard or inadequate information is available.

Appendix A: Full Resource Citations and Summaries

Article ID:	**151**
Citation:	NIOSH [2005]. International Chemical Safety Cards (ICSC): U.S. national version. [www.cdc.gov/niosh/ipcs/nicstart.html].
Resource type:	Web page
Educational materials:	No
Number of references:	
Industries/occupations:	General—overview
Specific process:	
Chemical:	Abrasives, cleaning agents, coolants, corrosives, heavy metals/inorganic compounds, organic dyes, particulates, pesticides, petroleum products & lubricants, plastics and resins, PCBs, solvents, other: comprehensive list of chemicals used in occupational settings
Specific chemicals:	There are currently cards for over 1500 chemicals.
Mixtures:	No
Audience:	General
Topics addressed:	C Hazard identification
	C.1 Risk phrases, hazard symbols, skin designations
	C.2 Tables/charts/lists of hazards for specific chemicals
	E Risk management
	E.1 Overview of skin exposure control options
	E.3 "Best practices"/guidelines/recommendations
	E.3.C Work practice/administration controls
	E.3.D PPE and PPE rules
	E.4 Guidelines/recommendations for postexposure skin decontamination
Summary:	ICSC Project is an undertaking of the International Programme on Chemical Safety (IPCS). The IPCS is a joint activity of three cooperating international organizations: the United Nations Environment Programme (UNEP), the ILO, and the WHO. Each ICSC summarizes essential health and safety information on chemicals for their use by workers and employers in factories, agriculture, construction, and other work places. They consist of a series of standard phrases, mainly summarizing health and safety information collected, verified and peer reviewed by internationally recognized experts, taking into account advice from manufacturers and poison control centers.

Indexed Dermal Bibliography

Appendix A: Full Resource Citations and Summaries

The U.S. national version of the ICSCs cited here has been modified by the National Institute for Occupational Safety and Health (NIOSH) to include the following:

- OSHA permissible exposure limits (PELs).
- NIOSH recommended exposure limits (RELs).
- Immediately dangerous to life and health values (IDLHs).
- Links to the Appendices in the *NIOSH Pocket Guide to Chemical Hazards*.

Each card briefly lists the routes of exposure, potential acute skin hazards, and symptoms for each specific chemical, as well as general prevention and first-aid measures.

Article ID:	**152**
Citation:	NIOSH [2005]. NIOSH pocket guide to chemical hazards. Cincinnati, OH: U.S. DHHS, PHS, CDC, NIOSH, DHHS (NIOSH) Publication No. 97-140.
Resource type:	Technical publication/report
Educational materials:	No
Number of references:	0
Industries/occupations:	
Specific process:	
Chemical:	Abrasives, cleaning agents, coolants, corrosives, fiberglass and other fibers, food products, hand cleansers, heavy metals/inorganic compounds, latex, nanoparticles, organic dyes, particulates, pesticides, petroleum products & lubricants, plastics and resins, PAHs, PCBs, rubber additives, soaps and detergents, solvents
Specific chemicals:	398 chemicals included
Mixtures:	No
Audience:	General
Topics addressed:	C Hazard identification
	C.2 Tables/charts/lists of hazards for specific chemicals
	E Risk management
	E.3 "Best practices"/guidelines/recommendations
	E.3.D PPE and PPE rules
	E.4 Guidelines/recommendations for postexposure skin decontamination

Appendix A: Full Resource Citations and Summaries

Summary:	The *NIOSH Pocket Guide to Chemical Hazards* is a source of general industrial hygiene information on several hundred chemicals/classes for workers, employers, and occupational health professionals. It provides exposure limits, exposure routes, respirator recommendations, PPE suggestions, and first aid for many of the 398 chemicals reviewed. It presents key information and data in abbreviated or tabular form for chemicals or substance groupings (e.g. cyanides, fluorides, manganese compounds) that are found in the work environment. This portable reference book helps in responding to workplace emergencies and preventing exposures to workers. It is designed to help users recognize and control occupational chemical hazards. It does not present data analyses.
	It is available online, on a CD, or as a hard copy, spiral bound document. It contains chemical-specific information on the skin designation (denoted as [skin]), which indicates the potential for dermal absorption. Skin exposure should be prevented as necessary through the use of good work practices and gloves, coveralls, goggles, and other appropriate equipment.

Article ID:	**153**	
Citation:	NIOSH [2005]. [www.cdc.gov/niosh/homepage.html].	
Resource type:	Web site	
Educational materials:	Yes	
Number of references:		
Industries/occupations:	General—overview, Agricultural, Cleaning/Janitorial/Maid, Service—Medical	
Specific process:		
Chemical:	General—overview, cleaning agents, coolants, corrosives, fiberglass and other fibers, heavy metals/inorganic compounds, latex, nanoparticles, organic dyes, particulates, pesticides, petroleum products & lubricants, plastics and resins, PAHs, PCBs, rubber additives, soaps and detergents, solvents	
Specific chemicals:		
Mixtures:	No	
Audience:	Professional	
Topics addressed:	A	Overview
	A.1	Occurrence of skin exposures in the workplace
	A.2	Health hazards resulting from skin exposure to chemicals

Indexed Dermal Bibliography

Appendix A: Full Resource Citations and Summaries

	A.3	Investigation, intervention, and control of occupational skin exposures
	A.5	Dermal regulations and skin notations
	B	Surveillance and clinical aspects
	B.1	Surveillance study reporting incidences of occupational skin exposures
	B.1.A	Skin exposure major focus
	B.2	Loss of workdays and impact on productivity
	C	Exposure characterization
	C.1	Workplace factors associated with harmful skin exposures
	C.2	Description of factors influencing exposure conditions
	C.2.A	Exposure intensity/frequency/duration
	C.2.B	Exposure concentration
	C.2.C	Skin area affected
	C.4	Direct methods to measure exposure
	C.4.A	Surfaces
	D	Hazard identification
	D.1	Potential health effects resulting from specific chemicals
	D.1.A	Irritant contact dermatitis
	D.1.B	Allergic contact dermatitis/sensitization
	D.1.C	Systemic toxicity
	F	Risk management
	F.1	Exposure control strategies
	F.1.A	Substitution
	F.1.B	Engineering controls
	F.1.C	Work practice/Administrative controls
	F.1.D	PPE and PPE rules
	F.1.E	Skin management, barrier creams, moisturizers, cleansers, and rubs
Summary:		NIOSH is the federal agency responsible for conducting research and making recommendations for the prevention of work-related injury and illness. NIOSH's Web site has a variety of Web pages with information on dermal exposure to chemicals. Many of these resources can be accessed through the NIOSH safety and health topic page, "Skin Exposures and Effects." This Web page contains links to many of the NIOSH resources on dermal exposure, including the NORA Dermal Exposure Research Program

Appendix A: Full Resource Citations and Summaries

(DERP), as well as updates on ongoing research and conferences. Information on dermal exposure can also be accessed by chemical or by industry and occupation and then examined for dermal exposure-related information.

Additional NIOSH resources on dermal exposure available from this Web site include:

- The Registry of Toxic Effects of Chemical Substances (RTECS) is a toxicological database of chemical data extracted from the open scientific literature. For each chemical, six types of toxicity data are included in the file: (1) primary irritation, (2) mutagenic effects, (3) reproductive effects, (4) tumorigenic effects, (5) acute toxicity, and (6) other multiple dose toxicity. Where available, it includes skin and eye irritation data. A subscription from one of RTECS' database vendors, listed on the Web page, is necessary to access the database.

- NIOSHTIC 2 is a searchable bibliographic database of occupational safety and health publications, documents, grant reports, and journal articles supported in whole or in part by NIOSH.

- The CPC Database can be searched by chemical, and for each chemical the user can find information on whether skin contact should be avoided and a list of recommended protective clothing barriers.

- International Chemical Safety Cards (ISCS) is a searchable database of basic health and safety information on (ultimately) 2000 chemicals and can be searched by chemical for potential dermal hazards.

- The *NIOSH Pocket Guide to Chemical Hazards* is a searchable source of general industrial hygiene information on several hundred chemicals/classes for workers, employers, and occupational health professionals, including information on routes of exposure, target organs, symptoms, and first-aid procedures.

- The Skin Permeation Calculator can be used to calculate the skin permeation coefficient (Kp), a measure of the conductance of skin to a particular chemical from a particular vehicle.

- The National Occupational Research Agenda (NORA) Allergic and Irritant Dermatitis (AID) Team. NORA, which is a framework to guide occupational safety and health research into the next decade, created the AID Team to promote research in this area.

Appendix A: Full Resource Citations and Summaries

- Proceedings of the International Conference on Occupational and Environmental Exposures of Skin to Chemicals: Science & Policy, September 2002.
- *Worker Health Chart Book 2004*: Chapter 2, "Fatal and Nonfatal Injuries, and Selected Illnesses and Conditions, Skin Diseases and Disorders," presents national surveillance data on skin diseases and disorders.
- *Occupational Dermatoses: A Program for Physicians* is a slide show that presents an overview of occupational dermatitis, including both surveillance data and photographs of different types of dermatitis.
- A NIOSH Alert on *Preventing Allergic Reactions to Natural Rubber Latex in the Workplace* is a comprehensive document that provides information on the recognition, evaluation, and control of exposure to natural latex products. It includes a list of a number of products found in the workplace that may contain latex.
- *Control of Exposure to Perchloroethylene in Commercial Dry Cleaning* is a guide which includes a description of methods that can be used for exposure control.

Article ID:	154
Citation:	National Library of Medicine (NLM) [2005]. Toxicology Data Network (TOXNET)—databases on toxicology, hazardous chemicals, environmental health, and toxic releases. [http://toxnet.nlm.nih.gov/].
Resource type:	Web site
Educational materials:	No
Number of references:	
Industries/occupations:	Agricultural, Beauty/Cosmetology, Cleaning/Janitorial/Maid, Construction, Forestry/Fisheries, Manufacturing—Chemical, Manufacturing—Other, Medical Services, Mining, Service—Food, Service—Medical, Service—Other, Transportation/Communications/Utility
Specific process:	Haz-Map is searchable by job name and job task
Chemical:	Abrasives, cleaning agents, coolants, corrosives, fiberglass and other fibers, heavy metals/inorganic compounds, latex, nanoparticles, organic dyes, particulates, pesticides, petroleum products & lubricants, plastics and resins, PAHs, PCBs, rubber additives, solvents
Specific chemicals:	

Appendix A: Full Resource Citations and Summaries

Mixtures:	No	
Audience:	Professional	
Topics addressed:	A	Overview
	A.1	Occurrence of skin exposures in the workplace
	A.2	Health hazards resulting from skin exposure to chemicals
	A.3	Investigation, intervention, and control of occupational skin exposures
	A.5	Dermal regulations and skin notations
	B	Surveillance and clinical aspects
	B.4	Clinical protocols for recognition of skin exposure health effects
	C	Exposure characterization
	C.1	Workplace factors associated with harmful skin exposures
	D	Hazard identification
	D.1	Potential health effects resulting from specific chemicals
	D.1.A	Irritant contact dermatitis
	D.1.B	Allergic contact dermatitis/sensitization
	D.1.C	Systemic toxicity
	D.1.D	Other health effects
	D.1.E	Contribution to overall exposure
	D.2	Summaries of health effects, dose-response relationships
	F	Risk management
	F.1	Exposure control strategies
	F.1.C	Work practice/Administrative controls
	F.1.D	PPE and PPE rules

Summary: TOXNET is a cluster of databases covering toxicology, hazardous chemicals, environmental health and related areas. It is managed by the Toxicology and Environmental Health Information Program (TEHIP) in the Division of Specialized Information Services (SIS) of the NLM. TOXNET provides free access to a variety toxicology databases, including those described below:

- Haz-Map: an occupational toxicology database designed primarily for health and safety professionals, but also for consumers seeking information about the health effects of exposure to approximately 1,000 chemicals and biological agents at work. Haz-Map links jobs and hazardous tasks with occupational

Appendix A: Full Resource Citations and Summaries

diseases and their symptoms (see ID 155 for more information). (http://hazmap.nlm.nih.gov/hazmapadv.html)

- HSDB: a comprehensive, peer-reviewed resource for toxicology information on over 4,900 potentially hazardous chemicals. HSDB also provides information on emergency handling procedures, industrial hygiene, environmental fate, human exposure, detection methods, and regulatory requirements. (http://toxnet.nlm.nih.gov/cgi-bin/sis/htmlgen?HSDB)

- TOXLINE: a bibliographic database providing comprehensive coverage of the biochemical, pharmacological, physiological, and toxicological effects of drugs and other chemicals from 1965 to the present.

- Integrated Risk Information System (IRIS): a database from the USEPA that contains health risk information on over 500 chemicals. IRIS risk assessment data have been scientifically reviewed by scientists and represents consensus.

- ChemIDplus: a database providing access to a variety of databases used for the identification of chemical substances cited in NLM databases. ChemIDplus is searchable by chemical name, synonym, CAS registry number, molecular formula, classification code, locator code, and structure. Links to available databases are provided. ChemIDplus contains over 379,000 chemical records, of which over 257,000 include chemical structures.

- Wireless Information System for Emergency Responders (WISER): a system designed to assist first responders in hazardous material incidents. It is available as a web-based, windows-based, or PDA application. It provides a wide range of information on hazardous substances, including substance identification support, physical characteristics, human health information, and containment and suppression advice.

Article ID:	**155**
Citation:	NLM [2005]. Hazardous Substances Data Bank (HSDB) [http://toxnet.nlm.nih.gov/cgi-bin/sis/htmlgen? HSDB].
Resource type:	Web page
Educational materials:	No

Appendix A: Full Resource Citations and Summaries

Number of references:		
Industries/occupations:	Agricultural, Beauty/Cosmetology, Cleaning/Janitorial/Maid, Construction, Forestry/Fisheries, Manufacturing—Chemical, Manufacturing—Medical Services, Mining, Service—Food, Service—Medical, Service—Transportation/Communications/Utility	
Specific process:		
Chemical:	Abrasives, cleaning agents, coolants, corrosives, fiberglass and other fibers, heavy metals/inorganic compounds, latex, nanoparticles, organic dyes, particulates, pesticides, petroleum products & lubricants, plastics and resins, PAHs, PCBs, rubber additives, solvents	
Specific chemicals:		
Mixtures:	No	
Audience:	Professional	
Topics addressed:	A	Overview
	A.5	Dermal regulations and skin notations
	B	Surveillance and clinical aspects
	B.1	Surveillance study reporting incidences of occupational skin exposures
	B.1.A	Skin exposure major focus
	B.1.B	Skin exposure minor focus
	C	Exposure characterization
	C.1	Workplace factors associated with harmful skin exposures
	D	Hazard identification
	D.1	Potential health effects resulting from specific chemicals
	D.1.A	Irritant contact dermatitis
	D.1.B	Allergic contact dermatitis/sensitization
	D.1.C	Systemic toxicity
	D.1.D	Other health effects
	F	Risk management
	F.1	Exposure control strategies
	F.1.C	Work practice/Administrative controls
	F.1.D	PPE and PPE rules
Summary:	The HSDB is a toxicology data file on the NLM TOXNET. It focuses on the toxicology of potentially hazardous chemicals. It includes information on human exposure, industrial hygiene, emergency handling procedures, environmental fate, regulatory requirements, and related	

Indexed Dermal Bibliography

areas. All data are referenced and derived from a core set of books, government documents, technical reports and selected primary journal literature. HSDB is peer reviewed by the Scientific Review Panel (SRP), a committee of experts in the major subject areas within HSDB's topic areas. HSDB is organized into individual chemical records and contains over 4,900 records. The following broad groupings of information are provided, if available, for each chemical:

- Human Health Effects
- Emergency Medical Treatment
- Animal Toxicity Studies
- Metabolism/Pharmacokinetics
- Pharmacology
- Environmental Fate/Exposure
- Chemical/Physical Properties
- Chemical Safety & Handling
- Occupational Exposure Standards
- Manufacturing/Use Information
- Laboratory Methods
- Special References
- Synonyms and Identifiers
- Administrative Information

Article ID:	156		
Citation:	National Ready Mixed Concrete Association (NRMCA) [2006]. [www.nrmca.org].		
Resource type:	Web site		
Educational materials:	Yes		
Number of references:			
Industries/occupations:	Construction		
Specific process:			
Chemical:	Corrosives		
Specific chemicals:	Portland cement		
Mixtures:	No		
Audience:	General		
Topics addressed:	A	Overview	
	A.1	Occurrence of skin exposures in the workplace	
	E	Risk management	

Appendix A: Full Resource Citations and Summaries

	E.1	Overview of skin exposure control options
	E.3	"Best practices"/guidelines/recommendations
	E.3.C	Work practice/administration controls
	E.3.D	PPE and PPE rules
	E.3.E	Skin management, barrier creams, moisturizers, cleansers, and rubs
Summary:		The NRMCA is an industrial organization for the ready-mix concrete industry. Through their Web site's E-Store, there are several resources available on dermal exposure hazards and controls associated with ready-mix concrete. These include:

- Cement Burn Awareness Kit (training material)
- Cement Burn Warning (Mini Poster)
- Safe Drum Cleaning (video)

Article ID:	**157**	
Citation:	Ness SA [1994]. Surface and dermal monitoring for toxic exposures. New York: Wiley & Sons.	
Resource type:	Book/monograph, whole	
Educational materials:	No	
Number of references:	1303	
Industries/occupations:	General—overview	
Specific process:		
Chemical:	General—overview, heavy metals/inorganic compounds, organic dyes, pesticides, PCBs, other: TCDD	
Specific chemicals:	Sampling methods listed for dozens of specific chemicals	
Mixtures:	No	
Audience:	Professional	
Topics addressed:	A	Overview
	A.2	Health hazards resulting from skin exposure to chemicals
	A.3	Investigation, intervention, and control of occupational skin exposures
	C	Exposure characterization
	C.2	Description of factors influencing exposure conditions
	C.2.A	Exposure intensity/frequency/duration
	C.2.B	Exposure concentration
	C.2.C	Skin area affected

Indexed Dermal Bibliography

Appendix A: Full Resource Citations and Summaries

C.2.E	Uptake
C.3	Checklists/questionnaires to quantify skin exposure incidences
C.4	Direct methods to measure exposure
C.4.A	Surfaces
C.4.B	Skin
F	Risk management
F.1	Exposure control strategies
F.1.D	PPE and PPE rules

Summary: This definitive book was one of the first comprehensive guides to surface and dermal sampling methods. This book is divided into four parts, with accompanying chapters in each part, as described below:

Part I—Chemical Hazards

    Ch. 1   Identifying Chemicals as Hazards
    Ch. 2   Factors Affecting Chemical Permeation
    Ch. 3   Chemical Protective Clothing

Part II—Developing Strategies for Sampling

    Ch. 4   Assessment of Workplace Exposures
    Ch. 5   Assessment of Community Exposures

Part III—Surface Monitoring

    Ch. 6   Introduction to Surface Monitoring
    Ch. 7   Surface Sampling for Chemicals
    Ch. 8   Surface Sampling for Microorganisms
    Ch. 9   Surface Sampling for Low Level Radiation
    Ch. 10 Decontamination

Part IV—Dermal Sampling Techniques

    Ch. 11 Introduction to Dermal Monitoring
    Ch. 12 Skin Sampling Methods, Part 1, Wiping Swabbing and Washing
    Ch. 13 Skin Sampling Methods, Part 2, Direct Reading
    Ch. 14 Pad Dosimetry Methods
    Ch. 15 Clothing for Dosimetry and Protection

The book also includes appendices on methods and studies of amines, metals, polychlorinated biphenyls, tetrachlorodibenzodioxins, and pesticides.

Article ID:	**158**
Citation:	Ness SA [2000]. Surface and dermal monitoring. In: Meyers RA, ed. Encyclopedia of analytical chemistry. New York: Wiley, pp. 4824–43.

Appendix A: Full Resource Citations and Summaries

Resource type:	Book/monograph, chapter
Educational materials:	No
Number of references:	61
Industries/occupations:	General—overview
Specific process:	
Chemical:	Fiberglass and other fibers, heavy metals/inorganic compounds, pesticides, PCBs
Specific chemicals:	
Mixtures:	No
Audience:	Professional
Topics addressed:	C Exposure characterization
	C.2 Description of factors influencing exposure conditions
	C.2.C Skin area affected
	C.2.E Uptake
	C.3 Checklists/questionnaires to quantify skin exposure incidences
	C.4 Direct methods to measure exposure
	C.4.A Surfaces
	C.4.B Skin
	C.5 Exposure modeling
Summary:	This chapter provides an overview of current methods used to perform surface and dermal monitoring for chemicals. It describes methods for measuring surface contamination, as well as discusses surface sampling media and sampling strategies. It also covers dermal monitoring methods which directly assess chemical contamination on a worker's skin or clothing. In addition, it contains a table listing guidelines and standards for surface sampling results for 20 different chemicals or chemical groups.

Article ID:	**159**
Citation:	Nielsen J B, Grandjean P [2004]. Criteria for skin notation in different countries. Am J Ind Med *45*(3):275–80.
Resource type:	Journal article—review, meta-analysis
Educational materials:	No
Number of references:	19
Industries/occupations:	
Specific process:	

Appendix A: Full Resource Citations and Summaries

Chemical:	Solvents
Specific chemicals:	Ethylamine, Cyanamide, Methacrylic acid, Sodium azide, Acroleine, Xylene, N-hexane, Toluene, Perchloroethylene, Benzene, 1,3,-butadiene, Ethylbenzene,
Mixtures:	No
Audience:	Professional
Topics addressed:	A Overview
	A.5 Dermal regulations and skin notations
Summary:	This paper compared skin notations on lists of exposure limits for industrial chemicals in six countries—the U.S., Netherlands, Denmark, Poland, Slovakia, and Germany—and found that one-third of industrial chemicals listed had a skin notation. Differences in criteria for assigning skin notations did not explain discrepancies between countries that otherwise had comparable occupational exposure limits (OELs).

Article ID:	**160**	
Citation:	Nygren O [2002]. New approaches for assessment of occupational exposure to metals using on-site measurements. J Environ Monit 4(5):623–27.	
Resource type:	Journal article—review, meta-analysis	
Educational materials:	No	
Number of references:	10	
Industries/occupations:		
Specific process:		
Chemical:	Heavy metals/inorganic compounds	
Specific chemicals:		
Mixtures:	No	
Audience:	Professional	
Topics addressed:	C	Exposure characterization
	C.4	Direct methods to measure exposure
	C.4.A	Surfaces
	C.4.B	Skin
Summary:	This article compares the accuracy of XRF florescent tracers to the traditional use of membrane filters followed by laboratory analysis. Tests were performed using dust, cobalt, nickel, and molydemum.	

Article ID:	**161**

Appendix A: Full Resource Citations and Summaries

Citation:	OSHA [2005]. Sampling for surface contamination. Washington, DC: U.S. DOL, OSHA, TED 1-0.15A.
Resource type:	Guideline
Educational materials:	No
Number of references:	7
Industries/occupations:	
Specific process:	
Chemical:	Plastics and resins, other: Isocyanates
Specific chemicals:	
Mixtures:	No
Audience:	Professional
Topics addressed:	A Overview
	A.5 Dermal regulations and skin notations
	C Exposure characterization
	C.4 Direct methods to measure exposure
	C.4.A Surfaces
	C.4.B Skin
Summary:	Section II, Chapter 2 of the *OSHA Technical Manual* describes surface sampling protocols for OSHA inspectors but is applicable to a wider audience. Substances with skin notations are listed in an appendix.

Article ID:	**162**
Citation:	OSHA [2005]. [www.osha.gov].
Resource type:	Web site
Educational materials:	No
Number of references:	
Industries/occupations:	General—overview, Agricultural, Cleaning/Janitorial/Maid, Construction, Manufacturing—Chemical, Manufacturing—Other, Medical Services, Mining, Service—Medical
Specific process:	
Chemical:	General—overview, coolants, corrosives, heavy metals/inorganic compounds, pesticides, petroleum products & lubricants, plastics and resins, solvents
Specific chemicals:	Acrylonitrile, benzene, 1, 3-butadiene, dry cleaning chemicals, CrVI, formaldehyde, isocyanates, methylene chloride, acrylonitrile, benzene, 1, 3-butadiene, dry cleaning chemicals, CrVI, formaldehyde, isocyanates, methylene chloride, among others.

Appendix A: Full Resource Citations and Summaries

Mixtures:	No	
Audience:	Professional	
Topics addressed:	A	Overview
	A.1	Occurrence of skin exposures in the workplace
	A.2	Health hazards resulting from skin exposure to chemicals
	A.5	Dermal regulations and skin notations
	C	Exposure characterization
	C.1	Workplace factors associated with harmful skin exposures
	C.2	Description of factors influencing exposure conditions
	C.2.A	Exposure intensity/frequency/duration
	C.2.B	Exposure concentration
	C.2.C	Skin area affected
	C.4	Direct methods to measure exposure
	C.4.A	Surfaces
	C.4.B	Skin
	C.4.C	Biomonitoring
	E	Risk assessment
	E.1	Guidelines for risk assessment or analysis
	E.1.A	Localized health effects
	E.1.B	Systemic health effects
	F	Risk management
	F.1	Exposure control strategies
	F.1.A	Substitution
	F.1.B	Engineering controls
	F.1.C	Work practice/Administrative controls
	F.1.D	PPE and PPE rules
	F.1.E	Skin management, barrier creams, moisturizers, cleansers, and rubs
Summary:	OSHA is the federal agency under the Department of Labor that publishes and enforces safety and health regulations for most businesses and industries in the United States. In recent years, OSHA's focus has been on enforcement as well as outreach, education, compliance assistance, and partnerships and cooperative programs. The OSHA Web site provides information and links to information on dermal exposure, including:	

Appendix A: Full Resource Citations and Summaries

- Health and Safety Topics: This Web page includes a link to OSHA's dermal exposure Web page as well as Web pages for specific chemicals with dermal exposure potential, such as acrylonitrile; benzene; 1, 3- butadiene; dry cleaning chemicals; CrVI; formaldehyde; isocyanates; methylene chloride; metalworking fluids; and solvents. There are also links to information on surface contamination associated with chemicals that have skin designations, and an up-to-date list of the OSHA standards that address dermal exposure. Additional topics include how to recognize hazardous dermal exposures, how to evaluate dermal exposures, and how to control dermal exposures including personal protective equipment.

- *OSHA Technical Manual*: This manual is used by OSHA compliance officers as a reference for technical information on occupational safety and health issues. It includes a number of chapters with information relevant to dermal exposure, including dermal exposure hazards specific to chemicals or processes, methods for sampling for surface contamination, chemical protective clothing guidelines, and a list of substances listed with skin notations or designations by ACGIH TLVs and/or OSHA PELs.

- *Evaluation Guidelines for Surface Sampling Methods*: A document developed to provide chemists with a uniform means for evaluating surface sampling methods with regards to sampling media, sampling techniques, and sample preparation for analysis.

- "Chemical Sampling Information": This Web page provides data on a large number of chemicals that may be encountered in industrial hygiene investigations. It is meant as a basic reference for OSHA personnel. For select chemicals it contains OSHA wipe sampling methods.

- OSHA standards: Dermal exposures are addressed in specific standards for general industry, shipyard employment, marine terminals, the construction industry and for the identification, classification, and regulation of carcinogens, in addition to being covered in Section 5(a)(1) of the OSH Act (the General Duty Clause) which require an employer to "furnish to each of his employees employment and a place of employment that is free from recognized hazards that are causing or are likely to cause death or serious physical harm to his employees." Below is a highlight of OSHA standards and directives (instructions for compliance officers). This is only a

Indexed Dermal Bibliography

partial list of references to skin exposure in OSHA standards, guidelines, and chemical sampling methods. For more and up-to-date information, see this Web site.

General Industry (29 CFR 1910)

1910 Subpart H, Hazardous materials
1910.120, Hazardous waste operations and emergency response
1910 Subpart I, Personal protective equipment
1910 Subpart Z, Toxic and hazardous substances
1910.1028, Benzene
1910.1044, 1,2-dibromo-3-chloropropane
1910.1045, Acrylonitrile
1910.1048, Formaldehyde
1910.1050, Methylenedianiline.
1910.1051, 1.3-Butadiene
1910.1052, Methylene chloride
1910.1200, Hazard communication

Shipyard Employment (29 CFR 1915)

1915 Subpart I, Personal protective equipment
1915 Subpart I Appendix A, Non-mandatory guidelines for hazard assessment, personal protective equipment (PPE) selection, and PPE training program

Marine Terminals (29 CFR 1917)

1917 Subpart B, Marine terminal operations
1917.28 Hazard communication

Construction (29 CFR 1926)

1926 Subpart D, Occupational health and environmental controls
1926.60, Methylenedianiline
1926.65, Hazardous waste operations and emergency response

Identification, Classification, and Regulation of Carcinogens (29 CFR 1990.103, Definitions)

Directives:

Enforcement Procedure for Occupational Exposure to Formaldehyde. Compliance directives (CPL) 02-02-052 [CPL 2-2.52], (1990, November 20).
Benzidine-Based Dyes: Direct Black 38, Direct Brown 95 and Direct Blue 6 Dyes. CPL 02-02-027 [CPL 2-2.27], (1980, February 22).

Article ID: **163**

Citation:	Oppl R, Kalberlah F, Evans PG, van Hemmen JJ. A toolkit for dermal risk assessment and management: An overview. Ann Occup Hyg 47(8):629–40.
Resource type:	Journal article—review, meta-analysis
Educational materials:	No
Number of references:	15
Industries/occupations:	
Specific process:	
Chemical:	
Specific chemicals:	
Mixtures:	No
Audience:	Professional
Topics addressed:	C Exposure characterization
	C.5 Exposure modeling
	E Risk assessment
	E.1 Guidelines for risk assessment or analysis
	E.1.A Localized health effects
	E.1.B Systemic health effects
	F Risk management
	F.2 Protocols for risk management
	F.2.A Development of exposure reduction goal (qualitative or quantitative)

Summary: This article is the 2nd article of a six-part series on RISKOFDERM, a tool for conducting risk assessments. The series was published in the Annals of Occupational Hygiene in 2003. The following briefly summarizes each paper in the series:

1. ID 212—Outlines a "toolkit" for conducting dermal occupational risk assessment.
2. ID 163—Describes the assumptions in the toolkit and describes an approach to exposure assessment used by the toolkit.
3. ID 139—Describes the determinants relevant for dermal exposure models in the scope of regulatory risk assessment.
4. ID 219—Describes how default dermal exposure values can be adjusted for specific work situations.
5. ID 100—Describes the derivation of the toolkit's default task-based dermal exposure values.
6. ID 193—Describes the development of "intrinsic toxicity" (IT) scores used for hazard characterization.

Appendix A: Full Resource Citations and Summaries

Article ID:	**164**
Citation:	Oregon Department of Human Services (ORDHS) [2005]. Oregon Worker Illness and Injury Prevention Program (OWIIPP). [http://oregon.gov/DHS/ph/owiipp/index.shtml].
Resource type:	Web site
Educational materials:	No
Number of references:	
Industries/occupations:	General—overview
Specific process:	
Chemical:	General—overview, latex, rubber additives
Specific chemicals:	
Mixtures:	No
Audience:	Professional
Topics addressed:	B — Surveillance and clinical aspects
	B.1 — Surveillance study reporting incidences of occupational skin exposures
	B.1.A — Skin exposure major focus
	B.2 — Loss of workdays and impact on productivity
	B.3 — Surveillance study protocols/procedures for gathering data
	F — Risk management
	F.1 — Exposure control strategies
	F.1.A — Substitution
	F.1.C — Work practice/Administrative controls
Summary:	The OWIIPP is working to identify and prevent targeted occupational illnesses and injuries in Oregonians. Workplace dermatitis and latex allergies are two of their targeted occupational illnesses.

Article ID:	**165**
Citation:	OR-OSHA [2006]. [www.cbs.state.or.us/external/osha/index.html].
Resource type:	Web site
Educational materials:	Yes
Number of references:	
Industries/occupations:	General—overview, Agricultural
Specific process:	
Chemical:	General—overview, latex, pesticides, PCBs
Specific chemicals:	

240 Indexed Dermal Bibliography

Appendix A: Full Resource Citations and Summaries

Mixtures:	No	
Audience:	General	
Topics addressed:	A	Overview
	A.2	Health hazards resulting from skin exposure to chemicals
	A.3	Dermal regulations and skin notations
	B	Exposure characterization
	B.1	Job/tasks, industries/processes, or chemicals associated with skin exposures
	B.2	Factors that influence exposure conditions
	B.2.A	Exposure intensity/frequency
	B.2.B	Exposure controls
	E	Risk management
	E.1	Overview of skin exposure control options
	E.3	"Best practices"/guidelines/recommendations
	E.3.C	Work practice/administration controls
	E.3.D	PPE and PPE rules
	E.3.E	Skin management, barrier creams, moisturizers, cleansers, and rubs

Summary: OR-OSHA, a division of the Oregon Department of Consumer and Business Services, enforces Oregon state workplace safety and health rules. Their Web site has a variety of resources, including pamphlets and brochures, on working safely with agricultural chemicals. Although these do not exclusively address skin exposure, they do address skin exposures in terms of overall routes of exposure to agricultural chemicals. Examples include:

- Safe practices when working around hazardous agricultural chemicals (brochure).
- Clothes washing for pesticide handlers (magnet).
- IH for the non-IH (workbook).
- Audio-visual library with several videos on skin exposure.
- Training materials for basic health and safety, including identifying and controlling hazards and personal protective equipment.

Article ID: 166

Citation: Organisation for Economic Co-operation and Development (OECD) [2005]. Guidelines for the testing

Appendix A: Full Resource Citations and Summaries

	of chemicals. [www.oecd.org/findDocument/0,2350, en_2649_34377_1_1_1_1_1,00.html].
Resource type:	Web page
Educational materials:	No
Number of references:	
Industries/occupations:	
Specific process:	
Chemical:	General—overview
Specific chemicals:	
Mixtures:	No
Audience:	Professional
Topics addressed:	D Hazard identification
	D.3 Characterization protocols
	D.3.A Corrosivity
	D.3.B Irritation potential
	D.3.C Sensitization potential
	D.3.D Potential to cause systemic effects
	D.3.E Measurement of skin permeation rates and reservoir effects
Summary:	The *Guidelines for the Testing of Chemicals* are a basic set of tools that are designed for use in regulatory safety testing, subsequent chemical product notification, and chemical registration. These are only guidelines. The existing guidelines are periodically updated and new guidelines are also added regularly. The dermal related guidelines listed below can be downloaded from the OECD site.

- OECD Guidelines for the Testing of Chemicals, Sections 1-5; Health Effects, Section 4
- 402 Acute Dermal Toxicity (updated guideline, adopted February 24, 1987)
- 404 Acute Dermal Irritation/Corrosion (updated guideline, adopted April 24, 2002)
- 406 Skin Sensitisation (updated guideline, adopted July 17, 1992)
- 410 Repeated Dose Dermal Toxicity: 21/28-day Study (original guideline, adopted May 12, 1981)
- 411 Subchronic Dermal Toxicity: 90-day Study (original guideline, adopted May 12, 1981)
- 427 Skin Absorption: In Vivo Method (original guideline, adopted April 13, 2004)

Appendix A: Full Resource Citations and Summaries

- 428 Skin Absorption: *In Vitro* Method (original guideline, adopted April 13, 2004)
- 429 Skin Sensitization: Local Lymph Node Assay (updated guideline, adopted April 24, 2002)
- 430 *In Vitro* Skin Corrosion: Transcutaneous Electrical Resistance Test (TER) (original guideline, adopted April 13, 2004)
- 431 *In Vitro* Skin Corrosion: Human Skin Model Test (original guideline, adopted April 13, 2004)
- 434 Acute Dermal Toxicity-Fixed Dose Procedure, Draft New Guideline (May 2004) (deadline for public comments passed: July 16, 2004)

Chemical Testing Guidelines

- No. 9: Guidance Document for the Conduct of Studies of Occupational Exposure to Pesticides During Agricultural Application
- No. 13: Detailed Review Document on Classification Systems for Sensitizing Substances in OECD Member Countries
- No. 16: Detailed Review Document on Classification Systems for Skin Irritation/Corrosion in OECD Member Countries

Article ID:	167
Citation:	OECD [2006]. OECD's database on chemical risk assessment models. [http://webdomino1.oecd.org/comnet/env/models.nsf].
Resource type:	Web page
Educational materials:	No
Number of references:	
Industries/occupations:	General—overview
Specific process:	
Chemical:	General—overview
Specific chemicals:	
Mixtures:	No
Audience:	Professional
Topics addressed:	C Exposure characterization
	C.5 Exposure modeling
	D Hazard identification

Appendix A: *Full Resource Citations and Summaries*

	D.3	Characterization protocols
	D.3.F	QSARs—development, validation, and application
Summary:	This searchable database includes information on models (computerized or capable of being computerized) that are used by OECD member governments and industry to predict health or environmental effects, exposure potential, and possible risks. The database can be searched by route of exposure and will list all dermal models. These are generic exposure models and not specific to occupational settings. The methods described here have not been evaluated or validated by OECD, and no endorsement of the methods by OECD should be inferred by the inclusion of certain methods in this Web page.	

Article ID:	168	
Citation:	Packham CL [1996]. Risk assessment and exposure control in an occupational setting. Curr Probl Dermatol *25*:133–44.	
Resource type:	Journal article—primary	
Educational materials:	No	
Number of references:	0	
Industries/occupations:		
Specific process:		
Chemical:		
Specific chemicals:		
Mixtures:	No	
Audience:	Professional	
Topics addressed:	E	Risk assessment
	E.1	Guidelines for risk assessment or analysis
	E.1.A	Localized health effects
	E.1.B	Systemic health effects
	E.2	Example of risk assessments
	F	Risk management
	F.1	Exposure control strategies
	F.1.B	Engineering controls
	F.1.C	Work practice/Administrative controls
	F.1.D	PPE and PPE rules
	F.2	Protocols for risk management
	F.2.B	Development of approach to achieve exposure reduction goal

Appendix A: Full Resource Citations and Summaries

Summary:	This paper presents a process by which managers can identify dermal risks of greatest concern, a necessary step prior to invoking risk management strategies.

Article ID:	**169**	
Citation:	Packham CL [1999]. Essentials of occupational skin management. Southport, UK: The Limited Edition Press.	
Resource type:	Book/monograph, whole	
Educational materials:	No	
Number of references:	95	
Industries/occupations:	General—overview, Agricultural, Beauty/Cosmetology, Cleaning/Janitorial/Maid, Construction, Manufacturing—Chemical, Manufacturing—Other, Service—Food, Service—Medical, Transportation/Communications/Utility, Other: aerospace, dentistry, pharmaceuticals, printing, textiles,	
Specific process:		
Chemical:	General—overview, plastics and resins, rubber additives	
Specific chemicals:		
Mixtures:	No	
Audience:	General	
Topics addressed:	A	Overview
	A.1	Occurrence of skin exposures in the workplace
	A.2	Health hazards resulting from skin exposure to chemicals
	A.3	Dermal regulations and skin notations
	B	Exposure characterization
	B.1	Job/tasks, industries/processes, or chemicals associated with skin exposures
	B.2	Factors that influence exposure conditions
	B.2.A	Exposure intensity/frequency
	B.2.B	Exposure controls
	C	Hazard identification
	C.3	Protocols/checklists to identify skin hazards in the workplace
	D	Risk assessment
	D.1	Protocols/checklists to identify exposure risk
	E	Risk management
	E.1	Overview of skin exposure control options

Appendix A: Full Resource Citations and Summaries

	E.2	Protocols/checklists to monitor potential exposures
	E.3	"Best practices"/guidelines/recommendations
	E.3.A	Substitution
	E.3.B	Engineering controls
	E.3.C	Work practice/administration controls
	E.3.D	PPE and PPE rules
	E.3.E	Skin management, barrier creams, moisturizers, cleansers, and rubs
	E.4	Guidelines/recommendations for postexposure skin decontamination

Summary: This comprehensive book combines elements of dermatology, occupational hygiene, and engineering and provides practical examples and solutions. It is clearly written and appears to be useful to both professionals and other users. Although the book is more focused on practical application rather than presenting scientific argument, it addresses many complex subjects, such as uptake, in a straightforward manner. Chapter headings are not self-evident, but the book contains an extensive index. Chapters are:

1. Dermatological Engineering
2. Legislation and the Skin at Work
3. The Skin as a Barrier
4. Occupational Skin Disease
5. Occupations and Occupational Skin Disease
6. Risk Assessment for Nonrespiratory Hazards
7. Exposure Control—Engineering
8. Exposure Control Through Protective Equipment
9. Selection and Use of Gloves
10. Barrier Creams—Myth or Magic Answer?
11. Skin Care
12. Cross Infection and the Skin
13. Creating an Effective Skin Management System
14. Investigating a Problem at Work
15. Technology and Skin Management

The author also has a Web site [www.enviroderm.co.uk] where this book and other resources reviewed in this indexed dermal bibliography can be purchased.

Appendix A: Full Resource Citations and Summaries

Article ID:	**170**
Citation:	Patlewicz G, Rodford R, Walker JD. Quantitative structure-activity relationships for predicting skin and eye irritation. Environ Toxicol Chem 22:1862–69.
Resource type:	Journal article—review, meta-analysis
Educational materials:	No
Number of references:	46
Industries/occupations:	
Specific process:	
Chemical:	
Specific chemicals:	
Mixtures:	No
Audience:	Professional
Topics addressed:	D Hazard identification
	D.3 Characterization protocols
	D.3.F QSARs—development, validation, and application
Summary:	This paper reviewed quantitative structure-activity relationships (QSARs) for predicting skin and eye irritation from existing experimental data.

Article ID:	**171**
Citation:	Paustenbach D, Leung HW, Rothrock JA [1999]. Health risk assessment. In: Adams RM, ed. Occupational skin disease. Philadelphia: Saunders, 291–323.
Resource type:	Book/monograph, chapter
Educational materials:	No
Number of references:	210
Industries/occupations:	
Specific process:	
Chemical:	
Specific chemicals:	
Mixtures:	Yes
Audience:	Professional
Topics addressed:	C Exposure characterization
	C.2 Description of factors influencing exposure conditions
	C.2.B Exposure concentration

Appendix A: *Full Resource Citations and Summaries*

	C.2.C	Skin area affected
	C.2.E	Uptake
	C.4	Direct methods to measure exposure
	C.4.A	Surfaces
	C.5	Exposure modeling
	D	Hazard identification
	D.2	Summaries of health effects, dose-response relationships
	D.3	Characterization protocols
	D.3.E	Measurement of skin permeation rates and reservoir effects
	E	Risk assessment
	E.1	Guidelines for risk assessment or analysis
	E.1.A	Localized health effects
	E.1.B	Systemic health effects
	E.2	Example of risk assessments
Summary:	This comprehensive reference by over 40 clinician-contributors discusses diagnosis, treatment, and prevention of occupational skin disease. This chapter addresses the four phases of risk assessment: hazard identification, dose-response assessment, exposure assessment, and risk characterization, as well as uptake, fate of chemicals, exposure pathways, models and modeling, exposure measuring, sensitization, and risk reduction.	

Article ID:	**172**	
Citation:	Phillips AM, Garrod AN [2001]. Assessment of dermal exposure—empirical models and indicative distributions. Appl Occup Environ Hyg *16*(2):323–28.	
Resource type:	Journal article—review, meta-analysis	
Educational materials:	No	
Number of references:	12	
Industries/occupations:		
Specific process:		
Chemical:		
Specific chemicals:		
Mixtures:	No	
Audience:	Professional	
Topics addressed:	C	Exposure characterization

Appendix A: Full Resource Citations and Summaries

	C.5	Exposure modeling
	E	Risk assessment
	E.2	Example of risk assessments
Summary:	This article by the U.K.'s HSE proposes an exposure assessment mechanism, the "indicative distribution approach," to use when little or no direct dermal exposure data are available. It allows one to conduct a risk assessment using a simple 12-box matrix based upon two variables: profile (narrow, medium, or wide geometric standard deviation) and deposition level (1–4 mg/minute).	

Article ID:	**173**	
Citation:	Poet TS [2000]. Assessing dermal absorption. Toxicol Sci 58(1):1–2.	
Resource type:	Journal article—review, meta-analysis	
Educational materials:	No	
Number of references:	7	
Industries/occupations:		
Specific process:		
Chemical:		
Specific chemicals:		
Mixtures:	No	
Audience:	Professional	
Topics addressed:	C	Exposure characterization
	C.2	Description of factors influencing exposure conditions
	C.2.E	Uptake
Summary:	This short article summarizes environmental and nonenvironmental factors that contribute to dermal absorption.	

Article ID:	**174**
Citation:	Portland Cement Association (PCA) [2006]. [www.cement.org/].
Resource type:	Web site
Educational materials:	Yes
Number of references:	
Industries/occupations:	Construction
Specific process:	

Appendix A: Full Resource Citations and Summaries

Chemical:	Corrosives
Specific chemicals:	Portland cement
Mixtures:	No
Audience:	General
Topics addressed:	A Overview
	A.1 Occurrence of skin exposures in the workplace
	A.2 Health hazards resulting from skin exposure to chemicals
	B Exposure characterization
	B.2 Factors that influence exposure conditions
	B.2.A Exposure intensity/frequency
	B.2.B Exposure controls
	E Risk management
	E.1 Overview of skin exposure control options
	E.3 "Best practices"/guidelines/recommendations
	E.3.C Work practice/administration controls
	E.3.D PPE and PPE rules
	E.3.E Skin management, barrier creams, moisturizers, cleansers, and rubs
Summary:	The Portland Cement Association represents cement companies in the United States and Canada. It provides, among other things, research, education, and public affairs programs. Resources on dermal exposure that are available through their Web site include a Web page on working safely with concrete, as well as the following publications:

- Skin Safety with Cement and Concrete (video and DVD)
- Working Safely with Concrete (brochure, downloadable from their Web site)

Article ID:	**175**
Citation:	Rietschel RL [2004]. Clues to an accurate diagnosis of contact dermatitis. Dermatol Ther *17*(3):224–30.
Resource type:	Journal article—review, meta-analysis
Educational materials:	No
Number of references:	23
Industries/occupations:	
Specific process:	
Chemical:	

Appendix A: Full Resource Citations and Summaries

Specific chemicals:		
Mixtures:	No	
Audience:	Professional	
Topics addressed:	B	Surveillance and clinical aspects
	B.1	Surveillance study reporting incidences of occupational skin exposures
	B.1.A	Skin exposure major focus
	C	Exposure characterization
	C.4	Direct methods to measure exposure
	C.4.B	Skin
	D	Hazard identification
	D.1	Potential health effects resulting from specific chemicals
	D.1.A	Irritant contact dermatitis
	D.1.B	Allergic contact dermatitis/sensitization
	D.2	Summaries of health effects, dose-response relationships
Summary:	This paper discusses historical, morphologic, and diagnostic steps one can take to accurately diagnosis contact dermatitis, both allergic and irritant. A comprehensive assessment of the patient's environment is needed to obtain a correct diagnosis.	

Article ID:	176
Citation:	Riviere JE [2002]. Percutaneous absorption of chemical mixtures relevant to the Gulf War. Raleigh, NC: North Carolina State University at Raleigh, ADA409100/XAB, -163.
Resource type:	Technical publication/report
Educational materials:	No
Number of references:	38
Industries/occupations:	U.S. Military
Specific process:	
Chemical:	Pesticides
Specific chemicals:	N,N-Diethyl-m-toluamide (DEET), permethrin, pyridostigmine bromide, iisopropylfluorphosphate (DFP), low-level sulfur mustard (RD), DFP, JP-8 jet fuel
Mixtures:	Yes
Audience:	Professional
Topics addressed:	C Exposure characterization

Appendix A: Full Resource Citations and Summaries

	C.2	Description of factors influencing exposure conditions
	C.2.E	Uptake
	D	Hazard identification
	D.3	Characterization protocols
	D.3.D	Potential to cause systemic effects
	D.3.E	Measurement of skin permeation rates and reservoir effects
Summary:	This report quantifies the dermal absorption and cutaneous toxicity of chemical mixtures associated with Gulf War Illness. The research focuses on how (14)C-permethrin, pyridostigmine bromide, diisopropylfluorphosphate (DFP), low-level sulfur mustard (RD), DFP, and JP-8 jet fuel affects exposure to N,N-Diethyl-m-toluamide (DEET). These data demonstrate an effect of systemic drugs on dermal absorption and underscore the complexity of risk assessments of complex chemical mixtures.	

Article ID:	177	
Citation:	Riviere JE [2004]. Quantitating absorption of complex chemical mixtures. Raleigh NC: North Carolina State University at Raleigh, College of Veterinary Medicine, PB2005-101509/XAB, -32. Resource type: Journal article—review, meta-analysis	
Educational materials:	No	
Number of references:	73	
Industries/occupations:		
Specific process:		
Chemical:	corrosives, pesticides	
Specific chemicals:	atrazine, chlorpyrifos, methylparathion, nonylphenol, pentachlorophenal, phenol, p-nitrophenyl, fenthion, propazine, simazine, triazine	
Mixtures:	Yes	
Audience:	Professional	
Topics addressed:	C	Exposure characterization
	C.2	Description of factors influencing exposure conditions
	C.2.E	Uptake
	D	Hazard identification
	D.3	Characterization protocols

	D.3.E	Measurement of skin permeation rates and reservoir effects
	E	Risk assessment
	E.2	Example of risk assessments
Summary:		Although exposure to complex mixtures of chemicals is typical, estimating exposure for risk assessment is difficult because available databases are based on single-chemical exposure. This paper presents the results of tests on chemical mixture interactions that affect percutaneous absorption to define the physical chemical characteristics of the mixture, and includes a discussion of several dermal models including QSPR and IPPSF.

Article ID:	**178**
Citation:	Riviere JE, Baynes RE, Smith C [2000]. Quantitating the percutaneous absorption of mechanistically defined chemical mixtures final report 15 Nov 1997–14 Nov 2000. Raleigh NC: North Carolina State University at Raleigh, Cutaneous Pharmacology and Toxicology Center, ADA386659/XAB, -109.
Resource type:	Technical publication/report
Educational materials:	No
Number of references:	7
Industries/occupations:	Transportation/Communications/Utility, Other: Jet aircraft
Specific process:	
Chemical:	Petroleum products & lubricants
Specific chemicals:	Jet fuels, Jet A, JP-8, JP-8 +100, jet fuel hydrocarbons, naphthalene, dodecane, hexadecane, jet fuel performance additives, DIEGME, 8Q2l, Stadis 450
Mixtures:	Yes
Audience:	Professional
Topics addressed:	

	C	Exposure characterization
	C.2	Description of factors influencing exposure conditions
	C.2.E	Uptake
	D	Hazard identification
	D.3	Characterization protocols
	D.3.E	Measurement of skin permeation rates and reservoir effects
Summary:		This report discusses the percutaneous absorption and cutaneous toxicity of jet fuels (Jet A, JP-8, JP-8 +100), jet fuel hydrocarbons (naphthalene, dodecane, hexadecane),

and performance additives (DIEGME, 8Q2l, Stadis 450). The report cites seven journal articles (authored or coauthored by Jim E. Riviere) that are included in full.

The articles are:

Riviere JE, Monteiro-Riviere NA, Brooks JD, Budsaba K, Smith CE [1999]. Dermal absorption and distribution of topically dosed jet fuels Jet A, JP-8, and JP-8(100). Toxicol Appl Pharmacol *160*:60–75.

Allen DG, Riviere JE, Monteiro-Riviere NA [2000]. Induction of early biomarkers of inflammation produced by keratinocytes exposed to jet fuels Jet-A, JP-8 and JP-8(100). J Biochem Molecular Toxicol *14*:231–237.

Budsaba K, Smith CE, Riviere JE [2000]. Compass plots: a combination of star plot and analysis of means (ANOM) to visualize significant interactions in complex toxicology studies. Toxicol Methods *10*:313–332.

Riviere JE, Brooks JD, and Qiao GL [2000]. Methods for assessing the percutaneous absorption of volatile chemicals in isolated perfused skin: studies with chloropentafluorobenzene (CPFB) and dichlorobenzene (DCB). Toxicol Methods *10*:265–281.

Baynes RE, Martin T, Craigmill AL, Riviere JE [1999]. Strategies for estimating provisional acceptable residues (PAR) for extra label drug use in livestock. Regulatory Toxicol Pharmacol *29*:287–299.

Allen DG, Riviere JE, Monteiro-Riviere NA [2001]. Cytokine induction as a measure of cutaneous toxicity in primary and immortalized porcine keratinocytes exposed to jet fuels and their relation to normal human keratinocytes. Toxicol Lett *119*:209–217.

Rhyne BN, Pirone JR, Monteiro-Riviere NA [2002]. The use of enzyme histochemistry in detecting cutaneous toxicity of three topically applied jet fuel mixtures. Toxicol Mechanisms and Methods *12*:17–34.

Article ID:	179
Citation:	Riviere JE, Monteiro-Riviere NA, Baynes RE, Xia X, Smith C [2004]. Quantitating the percutaneous absorption of mechanistically defined chemical mixtures final report 15 Dec 2000–14 Dec 2003. Raleigh NC: North Carolina State University at Raleigh, Cutaneous Pharmacology and Toxicology Center, ADA422081/XAB, -33.

Appendix A: Full Resource Citations and Summaries

Resource type:	Technical publication/report
Educational materials:	No
Number of references:	11
Industries/occupations:	Transportation/Communications/Utility, Other: Jet aircraft
Specific process:	
Chemical:	Petroleum products & lubricants
Specific chemicals:	Jet fuel, Jet A, JP-8, JP-8(100)
Mixtures:	No
Audience:	Professional

Topics addressed:

C	Exposure characterization
C.1	Workplace factors associated with harmful skin exposures
C.2	Description of factors influencing exposure conditions
C.2.E	Uptake
D	Hazard identification
D.1	Potential health effects resulting from specific chemicals
D.1.C	Systemic toxicity
D.2	Summaries of health effects, dose-response relationships

Summary: This report assesses the dermal absorption and skin toxicity of topically applied jet fuels and their additives using pigs, *in vitro* porcine skin and inert membrane models, as well as human keratinocyte cell cultures.

Article ID:	**180**
Citation:	Rodford R, Patlewicz G, Walker JD, Payne MP [2003]. Quantitative structure-activity relationships for predicting skin and respiratory sensitization. Environ Toxicol Chem 22(8):1855–61.
Resource type:	Journal article—review, meta-analysis
Educational materials:	No
Number of references:	28
Industries/occupations:	
Specific process:	
Chemical:	
Specific chemicals:	Solubility data provided for 15 chemicals
Mixtures:	No

Appendix A: Full Resource Citations and Summaries

Audience:	Professional	
Topics addressed:	D	Hazard identification
	D.1	Potential health effects resulting from specific chemicals
	D.1.B	Allergic contact dermatitis/sensitization
	D.3	Characterization protocols
	D.3.F	QSARs—development, validation, and application
Summary:	This paper reviewed quantitative structure-activity relationships (QSARs) for predicting skin and respiratory sensitization from existing experimental data.	

Article ID:	**181**	
Citation:	Romano-Woodward D [2000]. Safe use of glutaraldehyde. Nurs Stand *14*(32):47–51.	
Resource type:	Magazine article	
Educational materials:	No	
Number of references:	0	
Industries/occupations:		
Specific process:		
Chemical:	cleaning agents	
Specific chemicals:	glutaraldehyde	
Mixtures:	No	
Audience:	General	
Topics addressed:	A	Overview
	A.2	Health hazards resulting from skin exposure to chemicals
	E	Risk management
	E.2	Protocols/checklists to monitor potential exposures
	E.3	"Best practices"/guidelines/recommendations
	E.3.A	Substitution
	E.3.B	Engineering controls
	E.3.D	PPE and PPE rules
	E.4	Guidelines/recommendations for postexposure skin decontamination
Summary:	This article explains what precautions should be used for handling glutaraldehyde, a chemical used in many healthcare settings to sterilize instruments.	

Appendix A: Full Resource Citations and Summaries

Article ID:	**182**
Citation:	Ross JH, Dong MH, Krieger RI [2000]. Conservatism in pesticide exposure assessment. Regul Toxicol Pharmacol *31*(1):53–58.
Resource type:	Journal article—review, meta-analysis
Educational materials:	No
Number of references:	39
Industries/occupations:	
Specific process:	
Chemical:	
Specific chemicals:	
Mixtures:	No
Audience:	Professional
Topics addressed:	C Exposure characterization
	C.2 Description of factors influencing exposure conditions
	C.2.A Exposure intensity/frequency/duration
	C.2.E Uptake
	E Risk assessment
	E.1 Guidelines for risk assessment or analysis
	E.1.A Localized health effects
	E.1.B Systemic health effects
Summary:	This article discusses three exposure assessment factors that, could result in a significant overestimate of exposures to pesticides. The factors are (1) dermal absorption from animal studies, (2) daily dose extrapolated from partial day monitoring, and (3) nonbolus dosages from dermal or inhalation exposure. The authors recommend the generation of more appropriate data to minimize exposure overestimation, specifically human dermal absorption data, as well as conducting full-day exposure monitoring studies and, if feasible, generating dermal rather than oral toxicology data in those cases where the dermal route predominates.

Article ID:	**183**
Citation:	Rowse DH, Emmett EA [2004]. Solvents and the skin. Clin Occup Environ Med *4*(4):657–730, vi.
Resource type:	Journal article—review, meta-analysis
Educational materials:	No
Number of references:	331

Indexed Dermal Bibliography

Appendix A: Full Resource Citations and Summaries

Industries/occupations:	
Specific process:	
Chemical:	PAHs, solvents
Specific chemicals:	Table 1 provides applications, volatility, potential routes of entry, toxicity rating, skin lesions type, and health effects for 80 specific chemicals.
	Table 2 provides regulatory/guideline limits for 80 specific chemicals from the International Agency for Research on Cancer (IARC), ACGIH TWA, ACGIH short term exposure limit (STEL), ACGIH TLV, NIOSH, OSHA, and the Montreal protocol.
	Describes health effects, permeability, and other information on 40 specific alcohols, aldehydes, aliphatic and alicyclic hydrocarbons, amides, amines, aromatic hydrocarbons, chlorinated hydrocarbons, esters, ethers, glycol ethers, ketones, phenols, and terpenes.
Mixtures:	No
Audience:	Professional
Topics addressed:	A Overview
	A.2 Health hazards resulting from skin exposure to chemicals
	A.4 Skin physiology and function as barriers to chemical insults
	A.5 Dermal regulations and skin notations
	C Exposure characterization
	C.2 Description of factors influencing exposure conditions
	C.2.E Uptake
	D Hazard identification
	D.1 Potential health effects resulting from specific chemicals
	D.1.A Irritant contact dermatitis
	D.1.B Allergic contact dermatitis/sensitization
	D.1.C Systemic toxicity
	D.1.D Other health effects
	D.2 Summaries of health effects, dose-response relationships
	D.3 Characterization protocols
	D.3.E Measurement of skin permeation rates and reservoir effects
Summary:	This paper examines skin structure, permeability, and chemical uptake (injuries caused by solvents). Included are

reviews of solvent dermal health effects and the potential for systemic toxicity from dermal absorption.

Article ID:	**184**
Citation:	Roy A, Weisel CP, Lioy PJ, Georgopoulos PG [1996]. A distributed parameter physiologically-based pharmacokinetic model for dermal and inhalation exposure to volatile organic compounds. Risk Anal 16(2):147–60.
Resource type:	Journal article—review, meta-analysis
Educational materials:	No
Number of references:	30
Industries/occupations:	
Specific process:	
Chemical:	VOCs
Specific chemicals:	
Mixtures:	No
Audience:	Professional
Topics addressed:	C Exposure characterization
	C.2 Description of factors influencing exposure conditions
	C.2.A Exposure intensity/frequency/duration
	C.2.B Exposure concentration
	C.2.C Skin area affected
	C.2.E Uptake
	C.5 Exposure modeling
Summary:	This paper compares the way three models, developed to estimate dermal dose from exposures to toxic chemicals, estimate chloroform dose. Two are physiologically based pharmacokinetic models (PBPKs). The other is a more recently developed generalized "distributed parameter" physiologically based pharmacokinetic model (DP-PBPK), which has been developed for short-term exposures. The three models were evaluated by comparing simulated postexposure exhaled breath concentration profiles with measured concentrations following environmental chloroform exposures. All three models yielded estimates close to that of measured exhaled breath concentrations. Differences are described in detail.

Article ID:	**185**
Citation:	Sahmel J, Boeninger M. Dermal exposure assessments. In: Bullock W, et al., ed. A strategy for assessing and managing

Appendix A: Full Resource Citations and Summaries

                                              occupational exposures, 3rd ed. Fairfax, VA: AIHA Press, 137–61.

Resource type:	Book/monograph, chapter
Educational materials:	No
Number of references:	103
Industries/occupations:	
Specific process:	
Chemical:	
Specific chemicals:	
Mixtures:	No
Audience:	Professional
Topics addressed:	

	A	Overview
	A.4	Skin physiology and functions as a barrier to chemical insults
	C	Exposure characterization
	C.2	Description of factors influencing exposure conditions
	C.2.A	Exposure intensity/frequency/duration
	C.2.B	Exposure concentration
	C.2.C	Skin area affected
	C.2.E	Uptake
	C.4	Direct methods to measure exposure
	C.4.A	Surfaces
	C.4.B	Skin
	C.4.C	Biomonitoring
	C.5	Exposure modeling
	E	Risk assessment
	E.1	Guidelines for risk assessment or analysis
	E.1.A	Localized health effects
	E.1.B	Systemic health effects
	E.2	Example of risk assessments

Summary: Topics include

    Barriers to conducting dermal exposure assessments
    Evaluating dermal exposures for local and systemic toxicity
    Conducting dermal exposure assessments
    Characterizing dermal exposures
    Reviewing pertinent literature
    Toxicology and absorption data
    Dermal OELs

Appendix A: Full Resource Citations and Summaries

SEG determination for dermal exposures
Dermal exposure assessment factors
Dermal contact area
Dermal concentration
Dermal contract frequency
Dermal retention time
Dermal penetration potential
Affects of skin health on dermal penetration
Judging of dermal exposure profiles
Uncertain dermal exposures
Dermal modeling approaches
Dermal modeling approaches
Evaluation of quantitative or semiquantitative data
Example: Using skin monitoring data for lead in metal cutting
References

Article ID:	**186**
Citation:	Sarkis K [2000]. Protecting hands against chemical exposures. Occupational Hazards 62(8):53–56.
Resource type:	Magazine article
Educational materials:	Yes
Number of references:	0
Industries/occupations:	
Specific process:	
Chemical:	
Specific chemicals:	
Mixtures:	No
Audience:	General
Topics addressed:	E Risk management
	E.3 "Best practices"/guidelines/recommendations
	E.3.D PPE and PPE rules
Summary:	The paper discusses how gloves prevent skin exposures, as well as how to select gloves based upon the material being handled, the hazard involved, the task being performed, and comfort. It also discusses the pros and cons of latex, nitrile, neoprene, polyvinylchloride (PVC), polyvinylalcohol (PVA), butyl, and viton gloves.

Article ID:	**187**
Citation:	Sartorelli P, Andersen HR, Angerer J, Corish J, Drexler H, Goen T, Griffin P, Hotchkiss SA, Larese F, Montomoli L,

Appendix A: Full Resource Citations and Summaries

Perkins J, Schmelz M, van de Sandt J, Williams F [2000]. Percutaneous penetration studies for risk assessment. Environ Toxicol Pharmacol 8(2):133–52.

Resource type:	Journal article—review, meta-analysis
Educational materials:	No
Number of references:	77
Industries/occupations:	
Specific process:	
Chemical:	Soaps and detergents, solvents
Specific chemicals:	Dimethylsulfoxide (DMSO)
Mixtures:	No
Audience:	Professional

Topics addressed:

A	Overview
A.3	Investigation, intervention, and control of occupational skin exposures
C	Exposure characterization
C.2	Description of factors influencing exposure conditions
C.2.E	Uptake
D	Hazard identification
D.3	Characterization protocols
D.3.E	Measurement of skin permeation rates and reservoir effects
E	Risk assessment
E.1	Guidelines for risk assessment or analysis
E.1.A	Localized health effects

Summary: This paper by the Percutaneous Penetration Subgroup of the European Community's Dermal Exposure Network presents issues related to percutaneous penetration (uptake) rates for important chemicals, factors affecting those update rates, and gaps in knowledge in the field of percutaneous penetration. Sections include

1. Introduction
2. The use of percutaneous penetration data in risk assessment
3. Factor influencing the choice of cell characteristics for percutaneous penetration *in vitro* studies
4. Factors influencing the choice of the donor phase for percutaneous penetration *in vitro* studies

Appendix A: Full Resource Citations and Summaries

5. Factors influencing the choice of skin and membrane for percutaneous penetration *in vitro* studies
6. Factors influencing the choice of receptor fluids for percutaneous penetration *in vitro* studies
7. The presentation of *in vitro* percutaneous penetration results
8. Existing guidelines on percutaneous penetration *in vitro* studies
9. Prediction of plasma levels from penetration data
10. The influence of cutaneous metabolism on skin absorption
11. Criteria for the selection of reference compounds for *in vitro* percutaneous penetration
12. Correlation between *in vitro* and *in vivo* experiments
13. The use of microdialysis for the determination of dermal penetration of hazardous substances *in vivo*

Article ID:	188
Citation:	Sartorelli P [2002]. Dermal exposure assessment in occupational medicine. Occup Med (Lond) *52*(3):151–56.
Resource type:	Journal article—review, meta-analysis
Educational materials:	No
Number of references:	16
Industries/occupations:	
Specific process:	
Chemical:	
Specific chemicals:	
Mixtures:	No
Audience:	Professional
Topics addressed:	A Overview
	A.1 Occurrence of skin exposures in the workplace
	C Exposure characterization
	C.2 Description of factors influencing exposure conditions
	C.2.E Uptake
Summary:	This paper provides a discussion of various methods for assessing dermal exposure, including threshold limit

Appendix A: Full Resource Citations and Summaries

values, QSARs, and occupational exposure limit skin notations.

Article ID:	**189**
Citation:	Schlede E, Aberer W, Fuchs T, Gerner I, Lessmann H, Maurer T, Rossbacher R, Stropp G, Wagner E, Kayser D [2003]. Chemical substances and contact allergy—244 substances ranked according to allergenic potency. Toxicology *193*(3):219.
Resource type:	Journal article—review, meta-analysis
Educational materials:	No
Number of references:	15
Industries/occupations:	
Specific process:	
Chemical:	
Specific chemicals:	
Mixtures:	No
Audience:	Professional
Topics addressed:	D Hazard identification
	D.1 Potential health effects resulting from specific chemicals
	D.1.B Allergic contact dermatitis/sensitization
	D.3 Characterization protocols
	D.3.C Sensitization potential
Summary:	In 2001, 30 experts, including university dermatologists, industry representatives, and regulators, concluded a 15-year project to rank 244 chemicals by contact allergenic potency based on clinical and experimental data on humans and animals. The chemicals were assigned to one of three categories. Category A (98 substances) includes potent contact allergens with significant allergenic properties. Category B (77 substances) includes substances with a solid-based indication of a contact allergenic potential and substances with the capacity of cross-reactions. Category C (69 substances) includes substances with insignificant or questionable allergenic effects.

Article ID:	**190**
Citation:	Schneider T, Vermeulen R, Brouwer DH, Cherrie JW, Kromhout H, Fogh CL [1999]. Conceptual model for

Appendix A: Full Resource Citations and Summaries

	assessment of dermal exposure. Occup Environ Med 56(11):765–73.
Resource type:	Journal article—review, meta-analysis
Educational materials:	No
Number of references:	47
Industries/occupations:	
Specific process:	
Chemical:	
Specific chemicals:	
Mixtures:	No
Audience:	Professional
Topics addressed:	C Exposure characterization
	C.2 Description of factors influencing exposure conditions
	C.2.A Exposure intensity/frequency/duration
	C.2.B Exposure concentration
	C.2.C Skin area affected
	C.2.E Uptake
	C.4 Direct methods to measure exposure
	C.4.B Skin
	C.5 Exposure modeling
Summary:	This paper presents a multicompartment model for dermal exposure assessment. The model describes the transport of a contaminant mass from the source of the hazardous substance to the surface of the skin. The model also offers a standardized method of measurement using consistent terminology. The merits of existing models are also discussed.

Article ID:	**191**
Citation:	Schneider T, Cherrie JW, Vermeulen R, Kromhout H [2000]. Dermal exposure assessment. Ann Occup Hyg 44(7):493–99.
Resource type:	Journal article—review, meta-analysis
Educational materials:	No
Number of references:	35
Industries/occupations:	
Specific process:	
Chemical:	

Indexed Dermal Bibliography

Appendix A: Full Resource Citations and Summaries

Specific chemicals:	
Mixtures:	No
Audience:	Professional
Topics addressed:	C Exposure characterization
	C.4 Direct methods to measure exposure
	C.4.A Surfaces
	C.4.B Skin
	C.4.C Biomonitoring
	C.5 Exposure modeling
Summary:	The authors propose a theoretical strategy to assess dermal exposure based on a conceptual model for airborne contaminants. Many different skin and surface measurements are evaluated.

Article ID:	**192**
Citation:	Schnuch A, Lessmann H, Schulz KH, Becker D, Diepgen TL, Drexler H, Erdmann S, Fartasch M, Greim H, Kricke-Helling P, Merget R, Merk H, Nowak D, Rothe A, Stropp G, Uter W, Wallenstein G [2002]. When should a substance be designated as sensitizing for the skin ('Sh') or for the airways ('Sa')? Hum Exp Toxicol 21(8):439–44.
Resource type:	Journal article—review, meta-analysis
Educational materials:	No
Number of references:	10
Industries/occupations:	
Specific process:	
Chemical:	
Specific chemicals:	
Mixtures:	No
Audience:	Professional
Topics addressed:	A Overview
	A.5 Dermal regulations and skin notations
	D Hazard identification
	D.1 Potential health effects resulting from specific chemicals
	D.1.B Allergic contact dermatitis/sensitization
Summary:	The article reviews the criteria for determining when a substance should be deemed an airway sensitizer ("Sa") or skin sensitizer ("Sh") according to the list of maximum allowable concentration (MAK) and biological tolerance

Appendix A: Full Resource Citations and Summaries

(BAT) values published annually by the Commission of the Deutsche Forschungsgemeinschaft for the Investigation of Health Hazards of Chemical Compounds in the Work Area (MAK Commission). The authors conclude that MAK and BAT values make the classification of substances more rational, consistent, comprehensible, and transparent, but their application may not be necessary or possible in some cases.

Article ID:	**193**
Citation:	Schuhmacher-Wolz U, Kalberlah F, Oppl R, van Hemmen JJ [2003]. A toolkit for dermal risk assessment: Toxicological approach for hazard characterization. Ann Occup Hyg *47*(8):641–52.
Resource type:	Journal article—review, meta-analysis
Educational materials:	No
Number of references:	34
Industries/occupations:	
Specific process:	
Chemical:	
Specific chemicals:	
Mixtures:	No
Audience:	Professional
Topics addressed:	E Risk assessment
	E.1 Guidelines for risk assessment or analysis
	E.1.A Localized health effects
	E.1.B Systemic health effects
Summary:	This article is the 6th article of a 6-part series on RISKOFDERM, a tool for conducting risk assessments. The series was published in the Annals of Occupational Hygiene in 2003. The following briefly summarizes each paper in the series:

1. ID 212—Outlines a "toolkit" for conducting dermal occupational risk assessment.

2. ID 163—Describes the assumptions in the toolkit and describes an approach to exposure assessment used by the toolkit.

3. ID 139—Describes the determinants relevant for dermal exposure models in the scope of regulatory risk assessment.

4. ID 219—Describes how default dermal exposure values can be adjusted for specific work situations.

Appendix A: Full Resource Citations and Summaries

5. ID 100—Describes the derivation of the toolkit's default task-based dermal exposure values.
6. ID 193—Describes the development of "intrinsic toxicity" (IT) scores used for hazard characterization.

Article ID:	194
Citation:	Semple S [2004]. Dermal exposure to chemicals in the workplace: Just how important is skin absorption? Occup Environ Med 61(4):376–82.
Resource type:	Journal article—review, meta-analysis
Educational materials:	No
Number of references:	31
Industries/occupations:	General—overview
Specific process:	
Chemical:	General—overview, solvents
Specific chemicals:	
Mixtures:	No
Audience:	Professional
Topics addressed:	A Overview
	A.1 Occurrence of skin exposures in the workplace
	A.2 Health hazards resulting from skin exposure to chemicals
	A.4 Skin physiology and functions as a barrier to chemical insults
	C Exposure characterization
	C.2 Description of factors influencing exposure conditions
	C.2.A Exposure intensity/frequency/duration
	C.2.B Exposure concentration
	C.2.C Skin area affected
	C.2.E Uptake
	C.4 Direct methods to measure exposure
	C.4.B Skin
	C.4.C Biomonitoring
	C.5 Exposure modeling
Summary:	This paper discusses the importance of occupational dermal exposure, factors that influence exposure and absorption, and methods for measuring and assessing dermal exposure.

Appendix A: Full Resource Citations and Summaries

Article ID:	**195**
Citation:	Semple S, Brouwer DH, Dick F, Cherrie JW [2001]. A dermal model for spray painters, Part II: Estimating the deposition and uptake of solvents. Ann Occup Hyg 45(1):25–33.
Resource type:	Journal article—primary
Educational materials:	No
Number of references:	26
Industries/occupations:	Construction
Specific process:	Spray painting
Chemical:	
Specific chemicals:	
Mixtures:	No
Audience:	Professional

Topics addressed:		
	C	Exposure characterization
	C.1	Workplace factors associated with harmful skin exposures
	C.2	Description of factors influencing exposure conditions
	C.2.A	Exposure intensity/frequency/duration
	C.2.B	Exposure concentration
	C.2.C	Skin area affected
	C.2.E	Uptake

Summary: Part 2 of 2. This paper presents a model based upon "a process-based, structured approach" that both estimates occupational dermal exposure and uptake of solvents, using airless spray painters as an example. Estimates are based upon spray technique, object shape, workers' individual work practices, and droplet formation and deposition. Predicted exposure showed reasonable correlation with the actual measured exposure and the authors conclude that a structured, process-based approach has the potential to produce reliable estimates of dermal exposure. The authors also call for additional field studies.

Part 2 identifies the determinants of exposure, calculates the flux of solvent through the stratum corneum, and estimates total dermal uptake using a range of exposure scenarios.

Article ID:	**196**
Citation:	Shum KW, Meyer JD, Chen Y, Cherry N, Gawkrodger DJ [2003]. Occupational contact dermatitis to nickel:

Appendix A: Full Resource Citations and Summaries

	experience of the British dermatologists (EPIDERM) and occupational physicians (OPRA) surveillance schemes. Occup Environ Med *60*(12):954–57.
Resource type:	Journal article—primary
Educational materials:	No
Number of references:	31
Industries/occupations:	Beauty/Cosmetology, Service—Medical
Specific process:	The study focused on the following occupational categories:
	Hairdressers, Bar staff, Chefs/cooks, Retail cash and checkout operators, Catering assistants, Counter clerks/cashiers, Cleaners, Nurses, Metal workers, Sales assistants, Secretaries
Chemical:	Heavy metals/inorganic compounds
Specific chemicals:	Nickel
Mixtures:	No
Audience:	Professional
Topics addressed:	B Surveillance and clinical aspects
	B.1 Surveillance study reporting incidences of occupational skin exposures
	B.1.A Skin exposure major focus
Summary:	This study used occupational surveillance reporting databases (EPIDERM and OPRA) to determine to what extent nickel caused occupational contact dermatitis in the U.K. The study concluded that up to 12% of total estimated cases of occupational contact dermatitis were due in part to nickel exposure.

Article ID:	**197**
Citation:	Smith Pease CK [2003]. From xenobiotic chemistry and metabolism to better prediction and risk assessment of skin allergy. Toxicology *192*(1):1–22.
Resource type:	Journal article—review, meta-analysis
Educational materials:	No
Number of references:	63
Industries/occupations:	
Specific process:	
Chemical:	
Specific chemicals:	
Mixtures:	No

Audience:	Professional	
Topics addressed:	A	Overview
	A.4	Skin physiology and function as barriers to chemical insults
	C	Exposure characterization
	C.2	Description of factors influencing exposure conditions
	C.2.E	Uptake
	D	Hazard identification
	D.1	Potential health effects resulting from specific chemicals
	D.1.B	Allergic contact dermatitis/sensitization
	D.2	Summaries of health effects, dose-response relationships
	D.3	Characterization protocols
	D.3.C	Sensitization potential
	D.3.F	QSARs—development, validation, and application
Summary:	This review explores general chemical and metabolic processes involved in the process of skin sensitization to chemicals. It also discusses recent work using xenobiotics to explore sensitization mechanisms.	

Article ID:	198	
Citation:	Soutar A, Semple S, Aitken RJ, Robertson A [2000]. Use of patches and whole body sampling for the assessment of dermal exposure. Ann Occup Hyg 44(7):511–18.	
Resource type:	Journal article—review, meta-analysis	
Educational materials:	No	
Number of references:	37	
Industries/occupations:		
Specific process:		
Chemical:		
Specific chemicals:		
Mixtures:	No	
Audience:	Professional	
Topics addressed:	C	Exposure characterization
	C.4	Direct methods to measure exposure
	C.4.B	Skin

Appendix A: Full Resource Citations and Summaries

	C.5	Exposure modeling
Summary:	This paper details the principles underlying patch and whole body sampling and their advantages and disadvantages. This paper takes a recent conceptual model for dermal exposure and discusses the role that the various techniques may play in the application of this model.	

Article ID:	**199**
Citation:	Susitaival P, Flyvholm MA, Meding B, Kanerva L, Lindberg M, Svensson A, Olafsson JH [2003]. Nordic Occupational Skin Questionnaire (NOSQ-2002): a new tool for surveying occupational skin diseases and exposure. Contact Dermatitis 49(2):70–76
Resource type:	Journal article—review, meta-analysis
Educational materials:	No
Number of references:	46
Industries/occupations:	
Specific process:	
Chemical:	
Specific chemicals:	
Mixtures:	No
Audience:	Professional
Topics addressed:	B Surveillance and clinical aspects
	B.3 Surveillance study protocols/procedures for gathering data
Summary:	This paper describes the Nordic Occupational Skin Questionnaire (NOSQ-2002) (both short and long versions) for use in getting more survey data on the epidemiology of occupational skin diseases.

Article ID:	**200**
Citation:	Syracuse Research Corporation (SRC) [2006]. SRC business areas: Environmental science. [www.syrres.com/esc/default1.htm]. Date accessed: April 14, 2008.
Resource type:	Data file
Educational materials:	No
Number of references:	
Industries/occupations:	General—overview
Specific process:	

Appendix A: Full Resource Citations and Summaries

Chemical:	Abrasives, cleaning agents, coolants, corrosives, fiberglass and other fibers, heavy metals/inorganic compounds, organic dyes, particulates, pesticides, petroleum products & lubricants, plastics and resins, PAHs, PCBs, solvents
Specific chemicals:	
Mixtures:	No
Audience:	Professional

Topics addressed:

C	Exposure characterization
C.2	Description of factors influencing exposure conditions
C.2.E	Uptake
C.4	Direct methods to measure exposure
C.4.B	Skin
D	Hazard identification
D.1	Potential health effects resulting from specific chemicals
D.1.A	Irritant contact dermatitis
D.1.B	Allergic contact dermatitis/sensitization
D.3	Characterization protocols
D.3.D	Potential to cause systemic effects
D.3.E	Measurement of skin permeation rates and reservoir effects

Summary: SRC is a not-for-profit, independent, research and development organization. SRC's environmental science area has expertise in analyzing information on occupational and environmental hazards. SRC has made available a number of different software programs that relate to skin exposures to chemicals including:

- DermWin, which estimates the dermal permeability coefficient (Kp) used to estimate the potential for a chemical to be absorbed through the skin.
- KowWin, which estimates the log octanol-water partition coefficient, log P, of chemicals using an atom/fragment contribution method developed at SRC.
- WsKow, which estimates an octanol/water partition coefficient using the algorithms in SRC's LogKow Program and estimates a chemical's water solubility from this value.

SRC also developed the Toxic Substances Control Act Test Submission Database (TSCATS), which is used for the collection, maintenance, and dissemination of information on unpublished technical reports submitted by industry to the USEPA under TSCA.

Article ID:	**201**
Citation:	ten Berge W [2004]. Home page of Wil ten Berge: Wil ten Berge model for dermal absorption. [http://home.planet.nl/~wtberge/].
Resource type:	Web site
Educational materials:	No
Number of references:	
Industries/occupations:	
Specific process:	
Chemical:	
Specific chemicals:	
Mixtures:	No
Audience:	Professional
Topics addressed:	C Exposure characterization
	C.5 Exposure modeling
	D Hazard identification
	D.3 Characterization protocols
	D.3.F QSARs—development, validation, and application
Summary:	This home page of Wil ten Berge contains a downloadable version of the SKINDERM Program, which can to used for the estimation of the skin permeation coefficients of aqueous and vapor chemicals using physico-chemical properties of the chemical and the octanol/water partition coefficient. This program is based on a QSAR database developed by A. Wilschut, W.F. ten Berge, P.J. Robinson, and T.E. McKone in 1995. The program currently contains data for over 60 chemicals.

Article ID:	**202**
Citation:	Thrall KD, Poet TS, Corley RA, Tanojo H, Edwards JA, Weitz KK, Hui X, Maibach HI, Wester RC [2000]. A real-time in-vivo method for studying the percutaneous absorption of volatile chemicals. Int J Occup Environ Health 6(2):96.
Resource type:	Journal article—primary
Educational materials:	No

Appendix A: Full Resource Citations and Summaries

Number of references:	24
Industries/occupations:	
Specific process:	
Chemical:	Solvents
Specific chemicals:	Methyl chloroform, TCE, benzene
Mixtures:	No
Audience:	Professional

Topics addressed:

C	Exposure characterization
C.2	Description of factors influencing exposure conditions
C.2.E	Uptake
C.4	Direct methods to measure exposure
C.4.B	Skin
C.5	Exposure modeling
D	Hazard identification
D.2	Summaries of health effects, dose-response relationships
D.3	Characterization protocols
D.3.E	Measurement of skin permeation rates and reservoir effects

Summary: This paper presents estimates of percutaneous absorption of volatile chemicals. Dermal uptake of solvents under nonsteady-state conditions was determined using real-time breath analysis in rats, monkeys, and humans. Physiologically based pharmacokinetic (PBPK) models were used to estimate dermal permeability. The effects of the exposure matrix, occlusion versus nonocclusion, and species differences were compared for methyl chloroform, TCE, and benzene. The method was found to be sufficiently sensitive for animal and human dermal studies at low exposure concentrations over small body surface areas for short periods and using nonsteady-state exposure conditions.

Article ID:	**203**
Citation:	Toeppen-Sprigg B [1999]. Management of dermatitis in the rubber manufacturing industry. Occup Med 14(4):797–818.
Resource type:	Journal article—review, meta-analysis
Educational materials:	No
Number of references:	51

Indexed Dermal Bibliography

Appendix A: Full Resource Citations and Summaries

Industries/occupations:	Manufacturing—Rubber manufacturing
Specific process:	
Chemical:	Latex, rubber additives
Specific chemicals:	
Mixtures:	No
Audience:	Professional
Topics addressed:	A Overview
	A.1 Occurrence of skin exposures in the workplace
	A.2 Health hazards resulting from skin exposure to chemicals
	A.3 Investigation, intervention, and control of occupational skin exposures
	B Surveillance and clinical aspects
	B.1 Surveillance study reporting incidences of occupational skin exposures
	B.1.A Skin exposure major focus
	F Risk management
	F.1 Exposure control strategies
	F.1.A Substitution
	F.1.D PPE and PPE rules
	F.1.E Skin management, barrier creams, moisturizers, cleansers, and rubs
	F.2 Protocols for risk management
	F.2.B Development of approach to achieve exposure reduction goal
Summary:	This article examines dermatitis in the rubber industry. Contributing agents include both natural rubber and the various additives used in its manufacture. The paper reviews prevention and control measures such as substitution, PPE, barriers creams, and monitoring. It also includes a discussion of the diagnosis and treatment of dermatitis.

Article ID:	**204**
Citation:	Tupker RA [2003]. Prediction of irritancy in the human skin irritancy model and occupational setting. Contact Dermatitis *49*(2):61–69.
Resource type:	Journal article—review, meta-analysis
Educational materials:	No
Number of references:	64

Appendix A: Full Resource Citations and Summaries

Industries/occupations:		
Specific process:		
Chemical:		
Specific chemicals:		
Mixtures:	No	
Audience:	Professional	
Topics addressed:	A	Overview
	A.4	Skin physiology and function as barriers to chemical insults
	C	Exposure characterization
	C.2	Description of factors influencing exposure conditions
	C.2.A	Exposure intensity/frequency/duration
	C.2.B	Exposure concentration
	C.4	Direct methods to measure exposure
	C.4.B	Skin
Summary:	This review presents findings in the field of skin irritancy testing and discusses factors that determine irritancy testing outcome, including extrinsic and intrinsic factors such as prior exposure and genetics. This review also discusses the results from prospective cohort studies as they relate to factors influencing the development of occupational dermatitis.	

Article ID:	**205**	
Citation:	Unison [2005]. Health and safety zone. [www.unison.org.uk/safety/index.asp].	
Resource type:	Web page	
Educational materials:	No	
Number of references:		
Industries/occupations:	General—overview	
Specific process:		
Chemical:	General—overview	
Specific chemicals:		
Mixtures:	No	
Audience:	General	
Topics addressed:	A	Overview
	A.1	Occurrence of skin exposures in the workplace

Indexed Dermal Bibliography

Appendix A: Full Resource Citations and Summaries

	A.2	Health hazards resulting from skin exposure to chemicals
	B	Exposure characterization
	B.1	Job/tasks, industries/processes, or chemicals associated with skin exposures
	E	Risk management
	E.1	Overview of skin exposure control options
Summary:	UNISON is Great Britain's biggest trade union, with members working in the essential utilities and for private contractors providing public services. Their Web site includes a health and safety zone, which contains information on different health and safety topics. Included is an information sheet on dermatitis, which provides background information on causes of dermatitis in the workplace and steps that can be taken to prevent dermatitis. The information provided on legal requirements is specific to Great Britain.	

Article ID:	206	
Citation:	Oregon State University [2006]. University of California-Davis, Oregon State University, Michigan State University, Cornell University, and University of Idaho. The EXTension TOXicology NETwork (EXTOXNT). [http://extoxnet.orst.edu/ghindex.html].	
Resource type:	Web site	
Educational materials:	No	
Number of references:		
Industries/occupations:	Agricultural	
Specific process:		
Chemical:	Pesticides	
Specific chemicals:		
Mixtures:	No	
Audience:	Professional	
Topics addressed:	A	Overview
	A.2	Health hazards resulting from skin exposure to chemicals
	A.5	Dermal regulations and skin notations
	D	Hazard identification
	D.1	Potential health effects resulting from specific chemicals
	D.1.A	Irritant contact dermatitis

Appendix A: Full Resource Citations and Summaries

	D.1.B	Allergic contact dermatitis/sensitization
	D.1.C	Systemic toxicity
	D.2	Summaries of health effects, dose-response relationships
Summary:		EXTOXNET is a cooperative effort of University of California-Davis, Oregon State University, Michigan State University, Cornell University, and the University of Idaho. Their Web site contains resources about exposure to toxicants in our environment. The Pesticide Information Profiles contain basic toxicology information about pesticides, including, where relevant, information on skin exposure and toxicological effects.

Article ID:	**207**	
Citation:	U.S. Army Center for Health Promotion and Preventive Medicine (USACHPPM) [2006] [http://chppm-www.apgea.army.mil/].	
Resource type:	Web site	
Educational materials:	Yes	
Number of references:		
Industries/occupations:		
Specific process:	Military	
Chemical:	Pesticides, petroleum products & lubricants, solvents	
Specific chemicals:	JP-8 jet fuel, paints	
Mixtures:	No	
Audience:	General	
Topics addressed:	A	Overview
	A.1	Occurrence of skin exposures in the workplace
	B	Exposure characterization
	B.1	Job/tasks, industries/processes, or chemicals associated with skin exposures
	B.3	Protocols/checklists to characterize exposure to skin hazards
	E	Risk management
	E.3	"Best practices"/guidelines/recommendations
	E.3.C	Work practice/administration controls
	E.3.D	PPE and PPE rules
	E.3.E	Skin management, barrier creams, moisturizers, cleansers, and rubs

Appendix A: Full Resource Citations and Summaries

	E.4 Guidelines/recommendations for postexposure skin decontamination
Summary:	The U.S. Army Center for Health Promotion and Preventive Medicine's Web site has a variety of resources available about identifying, assessing and controlling environmental, occupational, and disease threats in support of the national military. Their Web site has Post Deployment Exposure Fact Sheets that address exposure, including dermal, to chemicals, such as paints, solvents, pesticides, and jet fuel. The fact sheets include information on personal protective equipment and control measures, signs and symptoms of acute and chronic exposure, reversibility of acute and chronic health effects, treatment options, and long term surveillance requirements.

Article ID:	208
Citation:	United States Department of Transportation (USDOT) [2004]. Emergency response guidebook. [http://hazmat.dot.gov/pubs/erg/gydebook.htm].
Resource type:	Web page
Educational materials:	No
Number of references:	
Industries/occupations:	General—overview
Specific process:	
Chemical:	General—overview, abrasives, cleaning agents, coolants, corrosives, fiberglass and other fibers, heavy metals/inorganic compounds, organic dyes, particulates, pesticides, petroleum products & lubricants, plastics and resins, PAHs, PCBs, rubber additives, solvents
Specific chemicals:	Covers hundreds of different chemicals
Mixtures:	No
Audience:	Professional
Topics addressed:	F Risk management
	F.1 Exposure control strategies
	F.1.A Substitution
	F.1.B Engineering controls
	F.1.C Work practice/Administrative controls
	F.1.D PPE and PPE rules
Summary:	The ERG (2004) was developed jointly by the USDOT, Transport Canada, and the Secretariat of Communications and Transportation of Mexico (SCT) for use by firefighters, police, and other emergency services personnel who may be the first to arrive at the scene of a transportation

Appendix A: Full Resource Citations and Summaries

incident involving a hazardous material. The ERG was designed as a guide to aid first responders in (1) quickly identifying the specific or generic classification of the material(s) involved in the incident and (2) protecting themselves and the general public during this initial response phase of the incident. The ERG is updated every three to four years.

Each chemical or material listed in the guide book is assigned a corresponding response guide number. The guides are then used to direct first responders on how to safely respond to hazardous material incidents. Information provided in the guides includes general health hazards including any associated dermal hazards, recommended personal protective equipment, and proper emergency response procedures. The guide can be searched by material name or identification number. The guidebook is available online as a searchable database as well as in hard copy form.

Article ID:	**209**
Citation:	USEPA [2005]. USEPA, OPPT: Exposure assessment tools and models. [www.epa.gov/opptintr/exposure/index.htm].
Resource type:	Web page
Educational materials:	No
Number of references:	
Industries/occupations:	General—overview
Specific process:	
Chemical:	General—overview
Specific chemicals:	
Mixtures:	No
Audience:	Professional
Topics addressed:	C Exposure characterization
	C.5 Exposure modeling
	E Risk assessment
	E.1 Guidelines for risk assessment or analysis
	E.1.A Localized health effects
	E.1.B Systemic health effects
	E.2 Example of risk assessments
Summary:	The USEPA OPPT has developed several exposure assessment methods, databases, and predictive models to help in evaluating, among other things, how workers may be exposed to chemicals. This Web page provides a table of the USEPA

Appendix A: Full Resource Citations and Summaries

exposure assessment tools and models, whether they include a dermal component, whether they address workplace exposures, and links to where they can be downloaded. Methods listed include exposure assessment screening tools.

Article ID:	**210**
Citation:	van de Sandt J, Johannes JM, Dellarco M, van Hemmen J. From dermal exposure to internal dose. J Exposure Sci Environ Epidemiol *17*:S38–S47.
Resource type:	Journal article—review, meta-analysis
Educational materials:	No
Number of references:	
Industries/occupations:	
Specific process:	
Chemical:	
Specific chemicals:	
Mixtures:	No
Audience:	Professional
Topics addressed:	A Overview
	A.4 Skin physiology and functions as a barrier to chemical insults
	C Exposure characterization
	C.4 Direct methods to measure exposure
	C.4.A Surfaces
	C.4.B Skin
	C.4.C Biomonitoring
	C.5 Exposure modeling
	D Hazard identification
	D.3 Characterization protocols
	D.3.E Measurement of skin permeation rates and reservoir effects
	D.3.F QSARs—development, validation, and application
	E Risk assessment
	E.2 Example of risk assessments
Summary:	This review article discusses risk assessment, methodologies to measure dermal exposure, bioavailability data, exposure conditions, and nontesting methods for skin absorption, such as QSAR.

Appendix A: Full Resource Citations and Summaries

Article ID:	**211**
Citation:	van Hemmen JJ, Brouwer DH. Assessment of dermal exposure to chemicals. Sci Total Environ 168(2):131–41.
Resource type:	Journal article—review, meta-analysis
Educational materials:	No
Number of references:	51
Industries/occupations:	
Specific process:	
Chemical:	
Specific chemicals:	
Mixtures:	No
Audience:	Professional
Topics addressed:	C Exposure characterization
	C.2 Description of factors influencing exposure conditions
	C.2.E Uptake
	C.4 Direct methods to measure exposure
	C.4.A Surfaces
	C.4.B Skin
	C.4.C Biomonitoring
Summary:	This paper compares qualitative, semiqualitative, and quantitative methods for assessing dermal exposure to chemicals. These methods include job (activity) exposure profiles, surrogate skin techniques, removal techniques, tracer techniques, biological monitoring, and surface sampling techniques. The paper compares the methods by validation tests, key input factors sampling area, and collection.

Article ID:	**212**
Citation:	van Hemmen J, Auffarth J, Evans PG, Rajan-Sithamparanadarajah B, Marquart H, Oppl R [2003]. RISKOFDERM: risk assessment of occupational dermal exposure to chemicals. An introduction to a series of papers on the development of a toolkit. Ann Occup Hyg 47(8):595–98.
Resource type:	Journal article—review, meta-analysis
Educational materials:	No
Number of references:	9
Industries/occupations:	

Indexed Dermal Bibliography

Appendix A: Full Resource Citations and Summaries

Specific process:	
Chemical:	
Specific chemicals:	
Mixtures:	No
Audience:	Professional
Topics addressed:	E Risk assessment
	E.1 Guidelines for risk assessment or analysis
	E.1.A Localized health effects
	E.1.B Systemic health effects
Summary:	This article is the 1st article of a 6-part series on RISKOFDERM, a tool for conducting risk assessments. The series was published in the Annals of Occupational Hygiene in 2003. The following briefly summarizes each paper in the series:

1. ID 212—Outlines a "toolkit" for conducting dermal occupational risk assessment.
2. ID 163—Describes the assumptions in the toolkit and describes an approach to exposure assessment used by the toolkit.
3. ID 139—Describes the determinants relevant for dermal exposure models in the scope of regulatory risk assessment.
4. ID 219—Describes how default dermal exposure values can be adjusted for specific work situations.
5. ID 100—Describes the derivation of the toolkit's default task-based dermal exposure values.
6. ID 193—Describes the development of "intrinsic toxicity" (IT) scores used for hazard characterization.

Article ID:	213
Citation:	van Rooij JGM, Jongeneelen FJ [2007]. Review of skin permeation hazard of bitumen fumes. J Occup Environ Hyg 4(S1):237–244.
Resource type:	Journal article—review, meta-analysis
Educational materials:	No
Number of references:	38
Industries/occupations:	Construction
Specific process:	Asphalt roofing, paving
Chemical:	Petroleum products & lubricants, PAHs
Specific chemicals:	Bitumen fumes

Appendix A: Full Resource Citations and Summaries

Mixtures:	No	
Audience:	Professional	
Topics addressed:	A	Overview
	A.1	Occurrence of skin exposures in the workplace
	A.2	Health hazards resulting from skin exposure to chemicals
	A.3	Investigation, intervention, and control of occupational skin exposures
	A.4	Skin physiology and functions as a barrier to chemical insults
	C	Exposure characterization
	C.2	Description of factors influencing exposure conditions
	C.2.E	Uptake
	D	Hazard identification
	D.1	Potential health effects resulting from specific chemicals
	D.1.D	Other health effects
Summary:	This study presents a summary of the literature regarding the skin permeation hazard of bitumen fumes among workers.	

Article ID:	**214**	
Citation:	van-Wendel-de-Joode B, Brouwer DH, Vermeulen R, van Hemmen JJ, Heederik D, Kromhout H [2003]. DREAM: A method for semi-quantitative dermal exposure assessment. Ann Occup Hyg 47(1):71–87.	
Resource type:	Journal article—primary	
Educational materials:	No	
Number of references:	42	
Industries/occupations:	Manufacturing—Other	
Specific process:		
Chemical:		
Specific chemicals:		
Mixtures:	No	
Audience:	Professional	
Topics addressed:	C	Exposure characterization
	C.5	Exposure modeling
Summary:	This paper describes the DeRmal Exposure AssessMent (DREAM) Model for accessing and evaluating occupational	

Indexed Dermal Bibliography

dermal exposure to chemical and biological agents. DREAM provides an initial assessment of dermal exposure levels to liquids and solids, a framework for measuring strategies, and a basis for implementing control strategies. Two examples from the car construction industry are discussed in detail.

Article ID:	**215**
Citation:	van-Wendel-de-Joode B, van Hemmen JJ, Meijster T, Major V, London L, Kromhout H [2005]. Reliability of a semi-quantitative method for dermal exposure assessment (DREAM). J Expo Anal Environ Epidemiol 15(1):111–20.
Resource type:	Journal article—primary
Educational materials:	No
Number of references:	21
Industries/occupations:	Agricultural, Manufacturing—Chemical, Transportation/Communications/Utility
Specific process:	Provides dermal risk calculations for 35 industrial and agricultural tasks.
Chemical:	
Specific chemicals:	
Mixtures:	No
Audience:	Professional

Topics addressed:

C	Exposure characterization
C.4	Direct methods to measure exposure
C.4.A	Surfaces
C.4.B	Skin
C.5	Exposure modeling

Summary: The reliability of DREAM, a semiquantitative dermal exposure assessment method, was assessed by using 29 observers (mainly occupational hygienists) who were asked to fill in DREAM while performing side-by-side observations for different tasks comprising dermal exposures to liquids, solids, and vapors. The authors concluded that DREAM is useful for estimating dermal exposure both for epidemiological research and for occupational hygiene practice.

Article ID:	**216**
Citation:	Vermeulen R, Stewart P, Kromhout H [2002]. Dermal exposure assessment in occupational epidemiologic research. Scand J Work Environ Health 28(6):371–85.

Appendix A: Full Resource Citations and Summaries

Resource type:	Journal article—review, meta-analysis
Educational materials:	No
Number of references:	110
Industries/occupations:	
Specific process:	
Chemical:	
Specific chemicals:	
Mixtures:	No
Audience:	Professional

Topics addressed:

C		Exposure characterization
C.2		Description of factors influencing exposure conditions
C.2.A		Exposure intensity/frequency/duration
C.2.B		Exposure concentration
C.2.C		Skin area affected
C.2.E		Uptake
C.5		Exposure modeling

Summary: This paper presents the results of a literature survey conducted to identify dermal exposure assessment methods. Variables discussed include intensity, frequency, and duration of exposure; the exposed surface area; and personal, temporal, and spatial variability in dermal exposure and uptake. Methods include qualitative, quantitative, and semiquantitative techniques. The paper focuses on dermal exposure assessment in relation to systemic effects, but local effects are also considered.

Article ID:	**217**
Citation:	Walker JD, Rodford R, Patlewicz G [2003]. Quantitative structure-activity relationships for predicting percutaneous absorption rates. Environ Toxicol Chem *22*(8):1870–84.
Resource type:	Journal article—review, meta-analysis
Educational materials:	No
Number of references:	50
Industries/occupations:	
Specific process:	
Chemical:	
Specific chemicals:	
Mixtures:	No

Indexed Dermal Bibliography

Appendix A: Full Resource Citations and Summaries

Audience:	Professional	
Topics addressed:	D	Hazard identification
	D.3	Characterization protocols
	D.3.F	QSAR—development, validation, and application
Summary:	This article reviews quantitative structure-activity relationships (QSAR) for predicting percutaneous absorption rates from existing experimental data. It also provide estimates on the number of workers exposed to 25 specific chemicals and permeability coefficients (Kp) for 83 chemicals.	

Article ID:	**218**	
Citation:	Wang RGM, Maibach H, Knaak B, eds. [1993]. Health risk assessment: Dermal and inhalation exposure and absorption of toxicants. Boca Raton, FL: CRC Press.	
Resource type:	Book/monograph, whole	
Educational materials:	No	
Number of references:	1718	
Industries/occupations:	General—overview	
Specific process:		
Chemical:	General—overview, solvents	
Specific chemicals:		
Mixtures:	No	
Audience:	Professional	
Topics addressed:	A	Overview
	A.4	Skin physiology and function as barriers to chemical insults
	C	Exposure characterization
	C.2	Description of factors influencing exposure conditions
	C.2.E	Uptake
	C.5	Exposure modeling
	D	Hazard identification
	D.1	Potential health effects resulting from specific chemicals
	D.1.D	Other health effects
	D.2	Summaries of health effects, dose-response relationships

Appendix A: Full Resource Citations and Summaries

Summary: This book, published by the California Environmental Protection Agency, has a focus on clinical issues. Information covered includes skin and inhalation exposure to toxicants, skin metabolism, absorption, pharmacokinetic modeling, dermal absorption, cholinesterase inhibition, adverse reproductive effects, carcinogenicity, PBPK modeling, cytochrome P-450 metabolism in skin, health effects, and the role of epidemiology in assessing the hazards of toxicology. The following is a list of the book's chapters:

Ch. 1   Physiologically Based Pharmacokinetic Modeling to Predict Tissue Dose and Cholinesterase Inhibition in Workers Exposed to Organophosphorus and Carbamate Pesticides

Ch. 2   The Application of Pharmacokinetic Models to Predict Target Dose

Ch. 3   Cytochrome P450-Dependent Metabolism of Drugs and Carcinogens in Skin

Ch. 4   Percutaneous Absorption

Ch. 5   *In Vitro* Skin Metabolism

Ch. 6   Animal Models for Percutaneous Absorption

Ch. 7   A Comparative Study of the Kinetics and Bioavailability of Pure and Soil-Absorbed Benzene, Toluene, and m-Xylene After Dermal Exposure

Ch. 8   Prediction of Human Percutaneous Absorption with Physicochemical Data

Ch. 9   Dermal Absorption of TCDD: Effect of Age

Ch. 10 Percutaneous Absorption of Chemicals From Water During Swimming and Bathing

Ch. 11 Percutaneous Absorption of Contaminants From Soil

Ch. 12 General Overview of Toxicological Responses and Routes of Chemical Exposure

Ch. 13 Acute Toxicity Testing by the Dermal Route

Ch. 14 Subchronic Dermal Exposure Studies With Industrial Chemicals

Ch. 15 The Dose Response of Percutaneous Absorption

Appendix A: Full Resource Citations and Summaries

Ch. 16 Reproductive and Developmental Toxicity Studies by Cutaneous Administration

Ch. 17 Dermal Carcinogenicity Studies of Petroleum-Derived Materials

Ch. 18 Comparison of Results from Carcinogenicity Tests of Two Halogenated Compounds by Oral, Dermal and Inhalation Routes

Ch. 19 The Objectives and Goals of Dermal Carcinogenicity Testing of Petroleum Liquids

Ch. 20 Chemical Carcinogenesis in Skin: Causation, Mechanism, and Role of Oncogenes

Ch. 21 Incorporating Biological Information into the Assessment of Cancer Risk to Humans Under Various Exposure Conditions and Issues Related to High Background Tumor Incidence Rates

Ch. 22 Phototoxicity of Topical and Systemic Agents

Ch. 23 Techniques for Assessing the Health Risks of Dermal Contact with Chemicals in the Environment

Ch. 24 Interspecies Extrapolation of Toxicological Data

Ch. 25 Human Skin Xenografts to Athymic Rodents as a System to Study Toxins Delivered to or Through the Skin

Ch. 26 The Isolated Perfused Porcine Skin Flap

Ch. 27 Perspectives on Assessment of Risk from Dermal Exposure to Polycyclic Aromatic Hydrocarbons

Ch. 28 The Paradox of Herbicide 2,4,-D Epidemiology

Ch. 29 A Review of Epidemiologic Studies with Regard to Routes of Exposure to Toxicant

Article ID:	219
Citation:	Warren N, Goede HA, Tijssen SC, Oppl R, Schipper HJ, van Hemmen JJ [2003]. Deriving default dermal exposure values for use in a risk assessment toolkit for small and medium-sized enterprises. Ann Occup Hyg 47(8):619–27.
Resource type:	Journal article—review, meta-analysis
Educational materials:	No
Number of references:	30

Appendix A: Full Resource Citations and Summaries

Industries/occupations:	General—overview
Specific process:	
Chemical:	General—overview
Specific chemicals:	
Mixtures:	No
Audience:	Professional
Topics addressed:	C Exposure characterization
	C.5 Exposure modeling
	E Risk assessment
	E.1 Guidelines for risk assessment or analysis
	E.1.A Localized health effects
	E.1.B Systemic health effects

Summary: This article is the 5th article of a 6-part series on RISKOFDERM, a tool for conducting risk assessments. The series was published in the Annals of Occupational Hygiene in 2003. The following briefly summarizes each paper in the series:

1. ID 212—Outlines a "toolkit" for conducting dermal occupational risk assessment.
2. ID 163—Describes the assumptions in the toolkit and describes an approach to exposure assessment used by the toolkit.
3. ID 139—Describes the determinants relevant for dermal exposure models in the scope of regulatory risk assessment.
4. ID 219—Describes how default dermal exposure values can be adjusted for specific work situations.
5. ID 100—Describes the derivation of the toolkit's default task-based dermal exposure values.
6. ID 193—Describes the development of "intrinsic toxicity" (IT) scores used for hazard characterization.

Article ID:	**220**
Citation:	Washington Department of Labor and Industry (WADLI). [2005]. Dermatitis: Safety and health assessment and research for prevention (SHARP). [www.lni.wa.gov/Safety/Research/OccHealth/Derm/default.asp#Resources].
Resource type:	Web page
Educational materials:	Yes
Number of references:	

Indexed Dermal Bibliography

Appendix A: Full Resource Citations and Summaries

Industries/occupations:	\multicolumn{2}{l}{General—overview, Agricultural, Manufacturing— Service—Medical}
Specific process:		
Chemical:	\multicolumn{2}{l}{General—overview, fiberglass and other fibers, latex, petroleum products & lubricants, plastics and resins, solvents}
Specific chemicals:		
Mixtures:	No	
Audience:	Professional	
Topics addressed:	A	Overview
	A.1	Occurrence of skin exposures in the workplace
	A.3	Investigation, intervention, and control of occupational skin exposures
	A.4	Skin physiology and function as barriers to chemical insults
	B	Surveillance and clinical aspects
	B.1	Surveillance study reporting incidences of occupational skin exposures
	B.1.A	Skin exposure major focus
	B.1.B	Skin exposure minor focus
	B.2	Loss of workdays and impact on productivity
	B.3	Surveillance study protocols/procedures for gathering data
	C	Exposure characterization
	C.1	Workplace factors associated with harmful skin exposures
	F	Risk management
	F.1	Exposure control strategies
	F.1.A	Substitution
	F.1.B	Engineering controls
	F.1.C	Work practice/Administrative controls
	F.1.D	PPE and PPE rules
	F.1.E	Skin management, barrier creams, moisturizers, cleansers, and rubs
Summary:	\multicolumn{2}{l}{The Safety and Health Assessment and Research for Prevention (SHARP) Program at the Washington State Department of Labor and Industries conducts work-related dermatitis research and surveillance. Under the Washington Sentinel Event Notification System for Occupational Risks (SENSOR) Dermatitis Program, SHARP has conducted surveillance on and prevention of work-related dermatitis. This Web site on skin disorders}

Appendix A: Full Resource Citations and Summaries

describes the research projects, educational materials and surveys produced by this project, as well as summaries of data collected. Examples of documents available on this Web site include:

- A guide to preventing dermatitis while working with advanced composite materials
- Metal Working Fluids: Prevention of skin problems when working with metal working fluids
- Clothing dermatitis and clothing-related skin conditions
- Skin health in agriculture
- Hand dermatitis in healthcare workers
- Prevention of hand dermatitis in the healthcare setting
- Latex sensitivity in Washington State acute care hospitals: A needs assessment and survey of awareness of the issues
- Latex sensitivity in Washington State acute care hospitals

Article ID:	**221**
Citation:	Weber LW [2003]. Development of occupational skin disease. In: DiNardi SR, ed. The occupational environment: its evaluation, control, and management, 2nd ed. Fairfax, VA: AIHA.
Resource type:	Book/monograph, chapter
Educational materials:	No
Number of references:	54
Industries/occupations:	
Specific process:	
Chemical:	
Specific chemicals:	
Mixtures:	No
Audience:	Professional
Topics addressed:	A Overview
	A.4 Skin physiology and functions as a barrier to chemical insults
	B Surveillance and clinical aspects
	B.4 Clinical protocols for recognition of skin exposure health effects

Appendix A: Full Resource Citations and Summaries

	C	Exposure characterization
	C.2	Description of factors influencing exposure conditions
	C.2.A	Exposure intensity/frequency/duration
	C.2.B	Exposure concentration
	C.2.C	Skin area affected
	C.2.E	Uptake
	D	Hazard identification
	D.1	Potential health effects resulting from specific chemicals
	D.1.A	Irritant contact dermatitis
	D.1.B	Allergic contact dermatitis/sensitization
Summary:		Chapter 18, "Development of Occupational Skin Disease," from the AIHA book, *The Occupational Environmental: Its Evaluation Control and Management (the White Book)*, gives a brief description of skin physiology and conditions that effect dermal exposure. It also discusses, for the industrial hygienist, medical evaluations of the skin for occupational skin disease.

Article ID:	**222**
Citation:	Wester RC, Maibach HI [2000]. Understanding percutaneous absorption for occupational health and safety. Int J Occup Environ Health 6(2):86–92.
Resource type:	Journal article—review, meta-analysis
Educational materials:	No
Number of references:	16
Industries/occupations:	
Specific process:	
Chemical:	General—overview, heavy metals/inorganic compounds, pesticides, PAHs, PCBs
Specific chemicals:	DDT Benzopyrene Chlordane Pentachlorophenol PCBs 2,4 D Arsenic Cadmium Mercury
Mixtures:	No

Appendix A: Full Resource Citations and Summaries

Audience:	Professional	
Topics addressed:	C	Exposure characterization
	C.2	Description of factors influencing exposure conditions
	C.2.E	Uptake
	C.5	Exposure modeling
Summary:	This paper describes percutaneous absorption, factors affecting absorption, and exposure monitoring methods. It also provides percutaneous absorption rates for several chemicals.	

Article ID:	**223**	
Citation:	Wigger-Alberti W, Elsner P [2003]. Occupational contact dermatitis in the textile industry. Curr Probl Dermatol *31*: 114–22.	
Resource type:	Journal article—review, meta-analysis	
Educational materials:	No	
Number of references:	68	
Industries/occupations:	Manufacturing—Textile	
Specific process:	Dyeing Finishing	
Chemical:	Organic dyes, plastics and resins	
Specific chemicals:	Formaldehyde	
Mixtures:	No	
Audience:	Professional	
Topics addressed:	B	Surveillance and clinical aspects
	B.1	Surveillance study reporting incidences of occupational skin exposures
	B.1.A	Skin exposure major focus
	D	Hazard identification
	D.1	Potential health effects resulting from specific chemicals
	D.1.A	Irritant contact dermatitis
	D.1.B	Allergic contact dermatitis/sensitization
Summary:	This paper discusses irritant and allergic contact dermatitis in the textile industry—primarily from resins, formaldehyde, and dyes—as well as tasks with exposure potential.	

Indexed Dermal Bibliography

Appendix A: Full Resource Citations and Summaries

Article ID:	**224**
Citation:	Winder C, Carmody M [2002]. The dermal toxicity of cement. Toxicol Ind Health *18*(7):321–31.
Resource type:	Journal article—review, meta-analysis
Educational materials:	No
Number of references:	102
Industries/occupations:	Construction
Specific process:	
Chemical:	Heavy metals/inorganic compounds, other: cement alkalines
Specific chemicals:	Chromium [III], chromium [VI] Lime (Anhydrous Calcium Hydroxide)
Mixtures:	No
Audience:	Professional
Topics addressed:	B Surveillance and clinical aspects
	B.1 Surveillance study reporting incidences of occupational skin exposures
	B.1.B Skin exposure minor focus
	B.2 Loss of workdays and impact on productivity
	D Hazard identification
	D.1 Potential health effects resulting from specific chemicals
	D.1.A Irritant contact dermatitis
	D.1.B Allergic contact dermatitis/sensitization
	D.1.D Other health effects
	F Risk management
	F.1 Exposure control strategies
	F.1.A Substitution
	F.1.C Work practice/Administrative controls
Summary:	Contact dermatitis is one of the most frequently reported health problems among construction workers. Cement's alkaline ingredients (such as lime) produce irritant contact dermatitis. Ingredients, such as chromium, produce allergic contact dermatitis. This paper lists steps to reduce exposures which have been proven to reduce allergic (but not irritant) dermatitis.
Article ID:	**225**
Citation:	World Health Organization (WHO) [2005]. International Programme on Chemical Safety (IPCS). [www.who.int/ipcs/en/].

Appendix A: Full Resource Citations and Summaries

Resource type:	Web site
Educational materials:	No
Number of references:	
Industries/occupations:	General—overview
Specific process:	
Chemical:	General—overview, cleaning agents, coolants, corrosives, heavy metals/inorganic compounds, pesticides, petroleum products & lubricants, plastics and resins, PAHs, PCBs, rubber additives, solvents
Specific chemicals:	Variety of chemicals included in chemical-specific hazard assessments.
Mixtures:	No
Audience:	Professional

Topics addressed:

B	Surveillance and clinical aspects
B.4	Clinical protocols for recognition of skin exposure health effects
D	Hazard identification
D.1	Potential health effects resulting from specific chemicals
D.1.A	Irritant contact dermatitis
D.1.B	Allergic contact dermatitis/sensitization
D.1.C	Systemic toxicity
D.1.D	Other health effects
D.1.E	Contribution to overall exposure
D.2	Summaries of health effects, dose-response relationships
D.3	Characterization protocols
D.3.E	Measurement of skin permeation rates and reservoir effects

Summary: The IPCS is a cooperative venture between WHO, UNEP, and ILO. The two main roles of the IPCS are to establish the scientific basis for the safe use of chemicals and to strengthen national capabilities and capacities for chemical safety. A variety of resources containing information on dermal exposures and exposures to chemicals in general can be found at this Web site. Information of interest includes:

- Concise International Chemical Assessment Documents (CICADs): reviews on the effects of over 60 chemicals on human health and the environment. Over a hundred chemicals are included. The CICADs characterize the hazard and dose-response

Appendix A: Full Resource Citations and Summaries

of exposure to chemicals and provide examples of exposure estimation and risk characterizations. Skin exposure information can be found under the occupational exposure section.

- IPCS INCHEM: access to peer-reviewed information on chemicals commonly used throughout the world and that occur as contaminants in the environment and food. IPCS INCHEM consolidates information from a number of different intergovernmental organizations.

- IPCS INTOX: a tool for poison centers and related units that provide information on preventing, evaluating, diagnosing, treating, and reporting on chemical emergencies.

- A glossary of exposure assessment-related terms: contains definitions for terms used in exposure assessment literature.

Article ID:	226	
Citation:	WHO [2005]. The IPCS: Environmental health criteria document on dermal absorption [Draft]. [www.who.int/ipcs/methods/dermal_absorption/en/].	
Resource type:	Technical publication/report	
Educational materials:	No	
Number of references:		
Industries/occupations:	General—overview	
Specific process:		
Chemical:	General—overview	
Specific chemicals:		
Mixtures:	No	
Audience:	Professional	
Topics addressed:	A	Overview
	A.4	Skin physiology and function as barriers to chemical insults
	C	Exposure characterization
	C.2	Description of factors influencing exposure conditions
	C.2.A	Exposure intensity/frequency/duration
	C.2.B	Exposure concentration
	C.2.C	Skin area affected
	C.2.E	Uptake

Appendix A: Full Resource Citations and Summaries

	D	Hazard identification
	D.3	Characterization protocols
	D.3.E	Measurement of skin permeation rates and reservoir effects
	D.3.F	QSARs—development, validation, and application
Summary:		This document provides an overview of percutaneous absorption of chemicals and the use of toxicokinetics data as part of the process of chemical risk assessment. It is not meant to be comprehensive, but rather to cover current topics of interest in the field.

Article ID:	227	
Citation:	Wu CF, Chiu HH [2007]. Rapid method for determining dermal exposures to pesticides by use of tape stripping and FTIR spectroscopy: A pilot study. J Occup Environ Hyg 4(12):952–58.	
Resource type:	Journal article—primary	
Educational materials:	No	
Number of references:	29	
Industries/occupations:		
Specific process:		
Chemical:	pesticides	
Specific chemicals:	chloropyrifos	
Mixtures:	No	
Audience:	Professional	
Topics addressed:	C	Exposure characterization
	C.4	Direct methods to measure exposure
	C.4.B	Skin
	C.6	Other
Summary:	This study ascertained the feasibility of using Fourier transform infrared spectroscopy (FTIR) to analyze tape-stripped samples to provide near real-time dermal exposure estimates. The feasibility of the stripping-FTIR approach was demonstrated.	

Article ID:	228
Citation:	Zeliger HI [2003]. Toxic effects of chemical mixtures. Arch Environ Health 58(1):23–29.

Appendix A: Full Resource Citations and Summaries

Resource type:	Journal article—review, meta-analysis
Educational materials:	No
Number of references:	47
Industries/occupations:	
Specific process:	
Chemical:	
Specific chemicals:	
Mixtures:	Yes
Audience:	Professional

Topics addressed:		
	C	Exposure characterization
	C.2	Description of factors influencing exposure conditions
	C.2.E	Uptake
	D	Hazard identification
	D.3	Characterization protocols
	D.3.E	Measurement of skin permeation rates and reservoir effects

Summary: Exposures to chemical mixtures have reportedly produced unexpected effects. Examination of new case studies, as well as those previously reported, shows that when the human body is exposed to mixtures of chemicals that include lipophilic and hydrophilic species, the lipophiles facilitate the absorption of the hydrophiles at enhanced levels and produce effects that are not expected from individual chemicals. These effects include enhanced acute and chronic responses, low-level concentration response, and unexpected target organ attack. Octanol:water partition coefficients are predictive of relative lipophilicity and hydrophilicity. The findings have implications for safe drinking water standards, air quality standards, safe industrial and environmental exposure levels, product formulation, product labeling, and protocols for toxicity testing of chemical products.

Article ID:	**229**
Citation:	Zhai H, Maibach HI [2004]. Dermatotoxicology, 6th ed. Boca Raton, FL: CRC Press.
Resource type:	Book/monograph, whole
Educational materials:	No
Number of references:	3518
Industries/occupations:	

Appendix A: Full Resource Citations and Summaries

Specific process:		
Chemical:		
Specific chemicals:		
Mixtures:	No	
Audience:	Professional	
Topics addressed:	A	Overview
	A.4	Skin physiology and function as barriers to chemical insults
	C	Exposure characterization
	C.2	Description of factors influencing exposure conditions
	C.2.E	Uptake
	C.5	Exposure modeling
	D	Hazard identification
	D.1	Potential health effects resulting from specific chemicals
	D.1.A	Irritant contact dermatitis
	D.1.B	Allergic contact dermatitis/sensitization
	D.1.C	Systemic toxicity
	D.1.D	Other health effects
	E	Risk assessment
	E.1	Guidelines for risk assessment or analysis
	E.1.A	Localized health effects
	E.1.B	Systemic health effects
	F	Risk management
	F.1	Exposure control strategies
	F.1.E	Skin management, barrier creams, moisturizers, cleansers, and rubs

Summary: *Dermatotoxicology, 6th edition* is a comprehensive reference book that includes information on the mechanisms of action of toxic substances on the skin, practical information on the various methods to evaluating dermal toxicity, and the latest developments in skin toxicology. The sixth edition contains 56 chapters, including a number of chapters covering factors influencing absorption and hazard characterization protocols, such as

1. Skin Permeability

3. Percutaneous Absorption of Complex Chemical Mixtures

4. Anatomical Factors Affecting Barrier Function

Appendix A: Full Resource Citations and Summaries

8. Sensitive Skin
11. Irritant Dermatitis (Irritation)
12. Allergic Contact Dermatitis
13. Irritant Contact Dermatitis Versus Allergic Contact Dermatitis
14. Molecular Basis of Allergic Contact Dermatitis
15. Systemic Contact Dermatitis
16. Permeability of Human Skin to Metals and Paths for Their Diffusion
27. Barrier Creams
29. Tape Stripping Method and Stratum Corneum
36. Animal, Human, and *In Vitro* Test Methods for Predicting Skin Irritation
38. Test Methods for Allergic Contact Dermatitis in Animals
39. Test Methods for Allergic Contact Dermatitis in Humans
52. Evaluating Efficacy of Barrier Creams: *In Vitro* and *In Vivo* Models
53. Light-Induced Dermal Toxicity: Effects on the Cellular and Molecular Level

www.ingramcontent.com/pod-product-compliance
Lightning Source LLC
Chambersburg PA
CBHW080235180526
45167CB00006B/2287